Artificial Intelligence in Phytochemistry

Yamin Bibi • Kulsoom Zahara
Adeeb Shehzad
Muhammad Jawad Khan • Marcello Iriti
Editors

Artificial Intelligence in Phytochemistry

A New Era of Drug Discovery

Springer

Editors
Yamin Bibi
Department of Botany
Rawalpindi Women University
Rawalpindi, Pakistan

Adeeb Shehzad
Biodiversity Research Centre
Dhofar University
Salalah, Oman

Marcello Iriti
Department of Biomedical, Surgical
and Dental Sciences
University of Milan
Milan, Italy

Kulsoom Zahara
Department of Botany
PMAS-Arid Agriculture University
Rawalpindi, Pakistan

Muhammad Jawad Khan
Department of Biomedical Engineering
and Sciences, School of Mechanical
and Manufacturing Engineering
National University
of Sciences and Technology
Islamabad, Pakistan

ISBN 978-981-92-0874-6 ISBN 978-981-92-0875-3 (eBook)
https://doi.org/10.1007/978-981-92-0875-3

This Springer imprint is published by the registered company Springer Nature Singapore Pte Ltd.
The registered company address is: 152 Beach Road, #21-01/04 Gateway East, Singapore 189721, Singapore

Preface

The integration of artificial intelligence (AI) into phytochemistry represents a profound transformation in the way natural products are explored, analyzed, and translated into therapeutic innovations. For centuries, plants have served as a cornerstone of drug discovery, offering a vast reservoir of bioactive compounds. Yet, the complexity of phytochemical diversity and the limitations of conventional screening methods have often slowed progress. Today, AI-driven approaches, ranging from machine learning algorithms to predictive modeling and network pharmacology, are accelerating this journey, enabling researchers to uncover novel insights with unprecedented speed and precision.

This book, *Artificial Intelligence in Phytochemistry: A New Era of Drug Discovery*, was conceived to highlight this paradigm shift. It brings together contributions from leading experts across disciplines, institutions, and countries, reflecting the truly global nature of this scientific endeavor. The chapters span fundamental concepts, computational methodologies, case studies, and future perspectives, offering readers a comprehensive view of how AI is reshaping phytochemical research and drug development pipelines.

Our aim is twofold: first, to provide a scholarly resource for researchers, educators, and practitioners seeking to understand and apply AI in natural product research; and second, to inspire new collaborations that bridge the gap between traditional phytochemistry and modern computational sciences. By presenting both theoretical frameworks and practical applications, this volume seeks to serve as a roadmap for advancing sustainable, innovative, and effective drug discovery strategies. We hope this book will encourage readers to embrace interdisciplinary thinking and harness the power of AI to unlock the therapeutic potential of nature.

Rawalpindi, Pakistan — Yamin Bibi
Rawalpindi, Pakistan — Kulsoom Zahara
Salalah, Oman — Adeeb Shehzad
Islamabad, Pakistan — Muhammad Jawad Khan
Milan, Italy — Marcello Iriti

Acknowledgments

We extend our heartfelt gratitude to all the contributing authors whose expertise, dedication, and scholarly rigor have shaped this book into a meaningful contribution to the evolving interface of artificial intelligence and phytochemistry. Their willingness to explore interdisciplinary frontiers and share original insights has been instrumental in realizing the vision of this book.

We are especially thankful to Springer Nature for their continued support, editorial guidance, and commitment to publishing innovative and impactful scientific work. Their professionalism and responsiveness throughout the process have been deeply appreciated.

We also acknowledge the institutions and research centers that supported the contributors in their work, including Rawalpindi Women University, Dhofar University, Prince Sattam bin Abdulaziz University, and the University of Milan. Their encouragement of collaborative and cross-disciplinary research has been vital to this endeavor.

Special thanks go to our peer reviewers and academic colleagues who provided critical feedback and helped refine the chapters to meet the highest standards of scientific integrity and clarity.

Where applicable, we disclose that some authors utilized non-generative AI tools (e.g., grammar checkers, reference managers) to assist in manuscript preparation. All content was reviewed and approved by the respective authors to ensure accuracy, originality, and compliance with ethical standards.

Finally, we are grateful to our families, mentors, and students for their patience, encouragement, and inspiration throughout this journey. This book is a testament to the power of collaboration, curiosity, and the shared pursuit of knowledge.

Contents

Part I Phytochemistry and Drug Discovery

1 **Foundation of Phytochemistry and AI-Driven Innovations in Drug Discovery** . . . 3
Nosheen Akhtar, Muhammad Azhar Jameel, Awais Ali, and Asim Shahzad
1.1 Introduction . . . 4
1.2 Historical Overview . . . 4
1.3 Origin of Phytochemistry . . . 6
1.4 Major Categories of Phytochemicals . . . 7
1.5 Techniques Used for Drug Discovery . . . 8
1.6 AI and Drug Discovery in the Modern Era . . . 9
1.6.1 Characterization of Molecular Structure . . . 11
1.6.2 Machine Learning . . . 11
1.6.3 AI in Action: Modelling for Phytochemicals . . . 13
1.7 Challenges and Suggestions . . . 14
1.8 Conclusion . . . 15
References . . . 15

Part II AI Simulation Techniques in Phytochemistry

2 **AI-Powered Prediction of Molecular Properties of Phytochemicals** . . . 19
Iqra Riaz, Shahid Ullah Khan, and Sohail
2.1 Introduction . . . 20
2.2 AI Methodologies for Property Prediction . . . 21
2.2.1 Neural Networks and Deep Learning . . . 21
2.2.2 Regression Models . . . 22
2.3 Real-Time Tools and Platforms . . . 23
2.3.1 DeepChem . . . 23
2.3.2 Chemprop . . . 24
2.4 Model Validation and Performance Metrics . . . 25
2.4.1 Cross-Validation Techniques . . . 25
2.4.2 Performance Metrics . . . 25
2.4.2.1 Regression Tasks . . . 25
2.4.2.2 Classification Tasks . . . 26

2.4.3 External Validation and Benchmarking 26
2.5 Challenges and Future Directions 26
2.5.1 Data Scarcity and Imbalance 26
2.5.2 Bias in Training Data 26
2.5.3 Emerging AI Solutions 27
2.5.3.1 Transfer Learning 27
2.5.3.2 Federated Learning 27
2.5.4 ADMET-AI Platform 28
2.5.5 Integration of Multi-omics and Systems Biology 28
2.6 Summary 29
References 29

3 Simulation of Drug-Phytochemical Interactions Using Artificial Intelligence (AI) 31
Rahees Zaheer, Maryam Aftab, Muhammad Israr Khan, and Shah Zareen
3.1 Introduction 32
3.2 Machine Learning Models for Drug–Phytochemical Binding Predictions 34
3.2.1 Machine Learning (ML) 34
3.2.1.1 Case Studies of ML-Based Models 36
3.2.1.2 Model Selection Concepts 36
3.2.2 Deep Learning in Phytochemical Research 37
3.2.3 AI in Phytochemical Research 39
3.2.3.1 Databases, Tools, and Evaluation Metrics 40
3.3 Molecular Docking Simulations with AI Enhancements 42
3.3.1 Molecular Dynamics (MD) 43
3.3.2 QSAR and Structure-Activity Relationship (SAR) in Dock 44
3.4 AI in Virtual Screening of Phytochemicals 44
3.4.1 Principles of Virtual Screening 45
3.4.2 QSAR Modeling in Virtual Screening 46
3.4.3 Deep Learning Pipelines for Screening 46
3.4.4 Integration of AI into Virtual Screening Workflows 46
3.4.5 Future Perspectives 47
References 47

4 AI-Driven Optimization of Phytochemical Drug Candidates 51
Sawaira Rahman, Saboon, and Sidra Amin
4.1 Introduction 52
4.2 Building the Foundational Data Universe: Beyond Simple Databases 53
4.3 In Silico Prioritization: From Virtual Screening to Mechanism-of-Action Prediction 53
4.4 De Novo Molecular Design: Generative AI for Lead Optimization 53

4.5 Predictive ADMET and Systems Pharmacology Integration 54
4.6 AI-Driven Molecular Optimization 54
4.6.1 The Core Challenge: From Discrete Molecules to a Differentiable Landscape 54
4.6.2 Advanced Paradigms: Reinforcement Learning and Policy Gradients 55
4.6.3 Applications of AI-Driven Molecular Optimization Using Gradient Descent Algorithms 56
4.7 Computational Toolkits for AI-Driven Optimization of Phytochemical Drug Candidates 56
4.8 Case Studies of Optimization of Phytochemical Drug Candidates Using AI-Directed Optimization 57
4.9 Drugs Design 58
4.9.1 Use Cases of Evolutionary Algorithms in Drug Discovery 59
4.9.1.1 De Novo Molecular Design and Scaffold Hopping 59
4.9.1.2 Multi-Objective Optimization of ADMET Properties 59
4.9:1.3 Structure-Based Drug Design and Molecular Docking 60
4.9.1.4 Optimizing Pharmacokinetic Profiles and Prodrug Design 61
4.9.1.5 Designing Focused Libraries for High-Throughput Screening 61
4.10 Tools of Evolutionary Algorithms in Drug Discovery 62
4.11 Workflow of Evolutionary Algorithms in Drug Discovery 63
4.12 Future Directions and Challenges 64
References 64

Part III Pharmacological Applications of AI in Phytopharmacology

5 AI Tools for Designing Phytopharmacological Agents 69
Muhammad Israr Khan and Gil Won Kim
5.1 Introduction 70
5.2 AI Platform for Plant-Based Drug Design 70
5.2.1 IBM Watson for Drug Discovery (WDD) Platform 71
5.2.2 Insilico Medicine 72
5.2.3 Atomwise (ATOMNEt) 73
5.2.4 MolAICal: A Computational Tool for 3D Structure-Based Drug Design Targeting Proteins 74
5.2.5 Benevolent AI 75
5.3 Integration of Natural Product Database 76
5.3.1 NPASS (Neonatal Pain, Agitation, and Sedation Scale) 76
5.3.2 Super Natural II: A Database of Natural Products 77
5.3.3 Server, Database, and System Requirements 77

5.4 Real-Time Applications and Integrated Threads 78
5.5 Conclusion 79
References. 79

6 Predicting Efficacy and Toxicity of Phytochemical Drugs with AI Models 83
Ambreen Najaf and Muhammad Umair Asghar
6.1 Introduction 84
6.2 Artificial Intelligence in Modern Drug Discovery. 86
6.3 Phytochemical-Based Drug Discovery: Opportunities and Challenges 88
6.4 AI Models for Predicting Efficacy of Phytochemicals 88
6.5 Integration of Multi-omics and Systems Biology with AI. 89
6.6 Explainable AI (XAI) and Model Interpretability 91
6.7 Limitations and Current Challenges 92
6.8 Future Perspectives 93
6.9 Conclusion 94
References. 95

7 Case Studies: AI in Preclinical and Clinical Phytopharmacology . . . 97
Nurhasni Hasan and Isha Shakoor
7.1 Introduction 98
7.2 Data Analysis in AI for Natural Products Discovery and Development 99
7.3 Case Studies. 101
7.3.1 *Withania somnifera* Metabolite Profiling Using AI-Powered Technology for the Preparation of Anti-inflammatory Medication 101
7.3.2 QSAR Modeling for Antioxidant Flavonoids 102
7.3.3 AI in Clinical Trial Design for Herbal Formulations. 103
7.3.4 Real-Time Monitoring of Herbal Drug Efficacy 103
7.3.5 Agentic AI in Pharma: Personalized Phytomedicine. 104
7.3.6 AI for Antibiotic Discovery from Natural Products (Halicin) 105
7.3.7 AI-Accelerated Cancer Drug Discovery from Traditional Chinese Medicine (TCM) 105
7.3.8 AI for Optimizing Extraction of Flavonoids (Quercetin and Kaempferol) 106
7.3.9 AI-Enhanced Clinical Trial Design and Patient Selection . . . 106
7.3.10 Predicting Herb/Drug Interactions Using Artificial Intelligence. 107
7.4 Use of Artificial Intelligence in Quality Control and Authentication of Herbal Products 107
7.4.1 Mining of Ethnopharmacological Knowledge Using Artificial Intelligence. 108
7.5 Conclusion 108
References. 108

Part IV Integration of AI in Phytopharmacological Research

8 AI Techniques in the Discovery of Novel Phytochemical Compounds 113
Rahmat Wali and Marcello Iriti
8.1 Introduction 114
8.2 AI-Powered Literature Mining and Ethnobotanical Intelligence... 114
8.2.1 NLP Foundations for Phytochemical Lead Mining 114
8.2.2 Large Language Models and Ontology-Driven Extraction... 115
8.2.3 Ethnobotanical Intelligence and Multimodal Extraction ... 116
8.2.4 Toward Comprehensive Phytochemical and Ethnobotanical Knowledge Graphs........ 117
8.3 Deep Learning for Spectral Pattern Recognition 118
8.3.1 AI-Assisted Interpretation of MS and NMR Datasets 118
8.3.2 Automated Peak Annotation, Clustering, and Mixture Analysis 119
8.3.3 Model Architectures: CNNs, Transformers, and End-to-End Learning........ 119
8.3.4 Model Architectures: CNNs, Transformers, and Sequence Learning 120
8.3.5 Integrating Spectral AI into Phytochemical Workflows.... 121
8.4 Generative AI in Novel Scaffold Design 122
8.4.1 Molecular Representations for Phytochemical Generation 122
8.4.2 VAEs, GANs, and Transformer Generators........ 122
8.4.3 Reinforcement Learning for Scaffold Ref 123
8.4.4 Role and Scope in Phytochemical Pip 124
8.5 AI in Untargeted Metabolomics and Compound Prioritization.... 124
8.5.1 High-Dimensional Profiling and Dimensionality Reduction 124
8.5.2 AI-Based Selection of Bioactive Candidates........ 125
8.5.3 Integration with Chemotaxonomic and Phylogenetic Data... 125
8.6 Graph-Based Discovery and Network Inference 126
8.6.1 Graph Representations in Phytochemical Discovery...... 126
8.6.2 Graph Neural Networks (GNNs) for Compound-Target and Pathway Inference........ 127
8.6.3 Mapping Phytochemical and Biosynthetic Networks 127
8.7 Multi-omics Fusion for Lead Identification........ 128
8.7.1 Integrating Transcriptomics, Proteomics, and Metabolomics........ 128
8.7.2 Identifying Biosynthetic Hotspots and Regulatory Nodes... 129
8.7.3 Case Examples and Practical Considerations 129
8.8 Challenges in AI-Driven Phytochemical Discovery 130
8.8.1 Data Sparsity and Heterogeneity 130
8.8.2 Model Bias, Overfitting: Implications for Interpretability in Complex Systems 131
8.8.3 Reproducibility, Benchmarking, and Standardization..... 131

8.9 Future Directions and Ethical Issues 131
8.9.1 Democratization of AI Technology for Biodiverse 131
8.9.2 Ethical Bioprospecting and Indigenous Knowledge 132
8.9.3 Open-Source and Collaborative Discovery 132
References 132

9 AI-Assisted Development of Personalized Phytopharmacological Therapies 135
Umme Ammara, Rina Agustina, and Qurat ul ain Sajid
9.1 Introduction 136
9.2 AI-Driven Patient Stratification 137
9.2.1 Genomic and Metabolomic Profiling for Patient Subgroup Identification 137
9.2.2 AI in Disease Diagnosis 138
9.2.3 AI in Personalized Herbal Medicine 139
9.3 Real-Time Application 140
9.3.1 Clinical Evidence of Ashwagandha in Stratified Therapies 140
9.3.2 AI-Stratified Randomized Trials for Enhanced Sensitivity 140
9.3.3 Proof of Concept-AI Re-analysis Rescues Stratified Subgroup Signals 141
9.3.4 Integrated Omics and Herbal Knowledge Platforms 141
9.4 Global Case Studies 143
9.4.1 Case Study 1: AI-Driven Herbal Matching in Europe 143
9.4.2 Case Study 2: AI in Herbal Safety and Standardization 143
9.4.3 Case Study 3: AI-Powered Formulation Research in India 144
9.5 Smart Delivery System 144
9.6 Emerging Trends 144
9.6.1 AI and Ethnobotany: Where Tradition Meets Modern Medicine 145
9.6.2 Natural Language Processing 146
9.6.3 Wearable Tech Integration: Real-Time Monitoring of Herbal Therapies 146
9.6.4 Generative AI in Formulation: Simulating Compound Interactions 147
9.7 Conclusion 147
References 148

10 Collaborative Strategies for Pharma Companies Using AI in Phytopharmacology 151
Iram Zaheer, Tayyaba Afzal, and Jaroslaw Prockow
10.1 Introduction 152
10.2 The Evolution of Collaboration in AI-Driven Pharmacology 152
10.2.1 Historical Context: From Data-Driven Discovery to Empirical Screening 152
10.2.2 Conceptual Change: Integrating AI into Phytochemical Knowledge Systems 153
10.2.3 Pharmaceutical Companies Collaboration 153
10.2.4 From Cooperation to Co-creation 154
10.3 Case Studies on Strategic AI Partnerships in Drug Discovery . . . 154
10.3.1 Merck KGaA and Exscientia: Generative AI for Molecular Optimization 155
10.3.2 Atomwise and Bayer: Deep Learning for Target Prediction 155
10.3.3 Emerging AI Applications in Natural Product Research. . . 156
10.4 Data-Sharing Models and Open-Source Platforms 156
10.4.1 Open Targets: AI-Powered Target Recognition 156
10.4.2 Machine Learning Ledger Orchestration for Drug Discovery (MELLODDY): Federated Learning for Private Information 157
10.4.3 Natural-Product Database Integration 158
10.4.4 Open Collaboration's Importance for Phytopharmacology 158
10.5 Collaborative AI-Phytopharmacology Conceptual Framework. . . 159
10.5.1 The Collaborative Ecosystem's Elements 159
10.5.2 The Innovation and Knowledge Flow 160
10.5.3 AI Integration Throughout the Phytopharmacology Process 161
10.5.4 Innovation Outcomes and Cross-Sector Synergy 161
10.6 Challenges/Difficulties, Ethics, and Future Prospects 161
10.6.1 Technical and Data Challenges 162
10.6.2 Ethical and Legal Considerations 162
10.6.3 Global Equity and Socioeconomic Issues 162
10.6.4 Prospects/Future Possibilities 163
References 163

Part V Challenges of Artificial Intelligence in Phytochemistry and Drug Discovery

11 Ethical Concerns in Leveraging AI for Phytochemistry 167
Naima Shahzadi, Raheem Shahzad, Ayesha Batool, Shifa Shaffique, Sobia Ikram, and Adeeb Shehzad
11.1 Introduction 168
11.1.1 Overview of Machine Learning and Deep Learning Approaches. 169
11.2 Integration of AI and IoT in Digital Agriculture 169
11.3 Advances in Computational Phytochemistry and Quantum Mechanical Approaches 170
11.4 Role of Analytical Techniques and Machine Learning in Phytochemical Characterization 170
11.5 Artificial Intelligence in Multi-Omics Integration and Phytochemical Discovery 171
11.6 Challenges and Limitations of AI in Botanical and Phytochemical Research 172
11.7 Ethical Concerns of AI. 172
11.7.1 Fairness 173
11.7.2 Algorithmic Bias Has the Potential to Negatively Impact Certain Social Groups 173
11.7.3 Data Privacy 174
11.7.4 Ethical Data Practice in AI Development 174
11.8 Workforce Training Challenges in the Era of AI-Driven Pharmaceuticals 174
11.9 Opportunities and Concerns of AI in the Future of Employment and Medicine 175
11.10 Summary 175
11.11 Future Recommendations 177
References. 177

12 Legal and Regulatory Implications of AI Utilization in Drug Development 179
Muhammad Salman Khalid, Raheem Shahzad, Luay Rashan, and Adeeb Shehzad
12.1 Introduction 180
12.2 AI Techniques in Phytochemical Analysis. 184
12.2.1 Machine Learning Algorithms for Chemical Profiling. . . . 184
12.2.2 Deep Learning for Structural Confirmation 184
12.2.3 Generative Models for Novel Metabolites 185
12.3 Legal Considerations in AI-Driven Drug Development. 186
12.3.1 Data Governance and Ownership. 186
12.3.1.1 Data Ownership 186
12.3.1.2 Data Licensing. 186
12.3.1.3 Data Localization Laws 188

12.4 Privacy and Data Protection . . . 189
12.4.1 Reidentification Risks . . . 189
12.4.2 GDPR and HIPAA Compliance . . . 190
12.4.3 Cross-Border Data Transfer . . . 190
12.5 Intellectual Property (IP) Challenges . . . 191
12.5.1 Inventorship of AI-Generated Biomolecules . . . 191
12.5.2 Protection of AI Models . . . 191
12.5.3 Freedom to Operate (FTO) . . . 192
12.5.4 Liability and Accountability . . . 192
12.5.5 Algorithmic Transparency and Explainability . . . 193
12.6 Regulatory Requirements for AI in Phytochemistry . . . 194
12.6.1 Good Machine Learning Practice (GMLP) . . . 194
12.6.2 Model Validation . . . 194
12.6.3 Reproducibility . . . 195
12.7 Clinical Development . . . 195
12.7.1 Regulatory Guidance for Trials . . . 195
12.7.2 Digital Health Technologies . . . 196
12.7.3 Post-Marketing Surveillance (PMS) . . . 196
12.7.4 Automated Signal Detection Risks . . . 197
12.7.5 RWE Integration . . . 197
12.8 Comparative Global Regulatory Landscape . . . 198
12.8.1 United States (FDA) . . . 198
12.8.2 European Union . . . 200
12.8.3 Nagoya Protocol on Access and Benefits Sharing (ABS) . . . 200
12.8.4 United Kingdom (MHRA) . . . 201
12.8.5 Asia-Pacific . . . 202
12.8.6 Other Global and Regional Policies . . . 203
12.9 Exploring AI-Enabled Privacy and Security Through Real-World Cases . . . 203
12.10 Ethical and Societal Considerations . . . 206
12.10.1 Ethical Use of Phytochemical Data . . . 206
12.10.2 Equity in AI-Enabled Phytochemistry . . . 206
12.10.3 Public Trust and Transparency . . . 207
12.11 Future Trends in Phytochemistry . . . 207
12.11.1 Standardization of AI Validation . . . 208
12.11.2 Continuous Monitoring and Adaptive Regulation . . . 208
12.11.3 Regulatory Sandboxes and Target Pilot Approach . . . 208
12.11.4 Strengthening IP Law for AI . . . 209
12.12 Conclusions . . . 209
References . . . 210

13 Managing Security Risks in AI Deployment 215
Nimra Mudassir, Asim Waris, and Adeeb Shehzad
13.1 Introduction 216
13.2 Conceptual Foundations................................ 217
13.2.1 Evolution of AI Security 217
13.2.2 Differences Between AI Security and Traditional Cybersecurity 218
13.2.3 AI System Lifecycle and Threat Modeling 218
13.3 Taxonomy of Security Risks in AI Deployment 219
13.3.1 Data- and Model-Related Risks.................... 219
13.3.2 Infrastructure, Operational, and Human Governance Risks 222
13.3.3 Systemic, Ethical, and Societal Risks 222
13.4 Why Are Traditional Security Measures Not Enough.......... 223
13.4.1 Limitations of Conventional Cybersecurity Controls ... 223
13.4.2 Unique Vulnerabilities in AI Pipelines and Attack Surfaces.................................... 223
13.4.3 Black-Box Models and Lack of Mature AI Security Standards 224
13.5 AI Risk Management Frameworks 224
13.5.1 Global and Industry Frameworks.................. 224
13.5.2 Comparative Analysis and Gaps 225
13.6 Secure AI Development Lifecycle (AI-SDLC) 225
13.6.1 Secure Design and Data Governance................ 225
13.6.2 Secure Model Training and Evaluation 226
13.6.3 Secure Deployment, MLOps Controls, and Post-Deployment Monitoring 226
13.7 Technical and Operational Controls......................... 226
13.7.1 Data Security Controls........................... 226
13.7.2 Model Hardening and Infrastructure Protection 227
13.7.3 Threat Detection and Incident Response in AI Systems ... 227
13.7.4 Monitoring and Detection 228
13.7.5 Incident Response and Recovery 228
13.8 Governance, Policies, and Organizational Controls 229
13.8.1 Human Oversight, Ethics, and Documentation 229
13.8.2 Compliance Requirements and Vendor Risk Management................................. 229
13.9 Case Studies of AI Security Failures 230
13.9.1 Real-World AI Security Failures in Critical Domains ... 230
13.9.2 Lessons Learned and Cross-Domain Patterns 230
13.10 Sector-Specific Security Considerations 231
13.10.1 Healthcare 231
13.10.2 Finance and Banking 231
13.10.3 Public Health, Defense, and Critical Infrastructure..... 232
13.10.4 Pharmaceutical R&D and Drug Discovery Security 232

13.11 Future Directions in AI Security 232
13.11.1 Evolving Red-Teaming and Responsible Scaling 233
13.11.2 Autonomous AI Agents and Alignment with Safety Engineering 233
13.11.3 Global Harmonization and Integration with Safety Standards 234
13.12 Conclusion 235
References 235

14 Strategic Solutions for Overcoming Challenges in AI Adoption 239
Tehreem Baig, Muhammad Jawad Khan, and Adeeb Shehzad
14.1 Introduction 240
14.2 AI in the Drug Development Lifecycle 241
14.2.1 Discovery and Target Identification 242
14.2.2 Molecular Characterization Learning 244
14.2.3 Prediction of Physicochemical Properties 244
14.2.4 Non-clinical Development and Safety Assessment 245
14.2.5 Clinical Trial Design, Conduct, and Analysis 246
14.2.6 Manufacturing and Quality Control 247
14.2.7 Post-authorization and Pharmacovigilance 248
14.3 Global Regulatory Landscape for AI in Drug Development 249
14.3.1 European Union, Europe 250
14.3.1.1 EMA 251
14.3.1.2 Food and Drug Administration (FDA), USA . . . 251
14.3.1.3 Regulatory Authorities, United Kingdom 252
14.3.2 Global Regulatory Convergence Challenges 253
14.4 Data Protection, Privacy, and Data Governance 253
14.4.1 Nature of Data Used by AI in Drug Development 253
14.4.2 Lawful Bases for Processing and Secondary Use 254
14.5 Governance, Equity, and Future Directions for AI in Drug Development 255
14.6 Conclusion 257
References 257

Editors and Contributors

About the Editors

Yamin Bibi is Chairperson of the Department of Botany at Rawalpindi Women University, Pakistan. Her research focuses on plant-based natural products, particularly the isolation and characterization of bioactive compounds and their pharmacological evaluation. She has supervised numerous PhD and MPhil students, completed funded projects, and published widely in national and international journals. A recipient of multiple research awards, she brings her expertise in natural product chemistry and leadership as Corresponding Editor to highlight how artificial intelligence can accelerate discovery of plant-derived therapeutics.

Kulsoom Zahara is an emerging scholar in phytochemistry and pharmacognosy. She specializes in isolating and identifying bioactive compounds using advanced techniques such as HPLC-DAD and GC-MS. Her research explores anticancer and antimicrobial properties of diverse plants, with growing emphasis on in silico and AI-driven approaches to predict bioactivity and safety. She contributes expertise in plant-based drug discovery and food safety, underscoring the transformative role of AI in modern phytochemical research.

Adeeb Shehzad is an Associate Professor at the Biodiversity Research Center, Dhofar University, Oman. A globally recognized biomedical scientist, his pioneering work spans molecular biology, nanomedicine, and translational therapeutics. With advanced degrees from Kyungpook National University, South Korea, and prior appointments in Pakistan and Saudi Arabia, he has published over 100 papers with an h-index of 34. His research on cell signaling, nanomaterials, and regenerative medicine bridges fundamental science with clinical innovation. Listed among the World's Top 2% Scientists, he continues to advance biomedical education and sustainable health innovation.

Muhammad Jawad Khan is an Associate Professor at Prince Sattam bin Abdulaziz University, Saudi Arabia. He specializes in artificial intelligence, disease modeling, and biomedical diagnostics, with a focus on neurological disorders. Holding a PhD in Mechanical Engineering from Pusan National University, South Korea, he has advanced hybrid EEG-fNIRS systems for neuro-robotics and medical diagnostics. His interdisciplinary work spans AI-driven modeling, robotics, and prosthetic limb design. Recognized among the World's Top 2% Researchers (Stanford, 2022), he has secured major research funding and led award-winning projects such as the iMeche UAS Challenge 2022.

Marcello Iriti is Professor of Plant Biology and Pathology at the University of Milan, Italy. His research covers phytochemistry, pharmacology, nutraceuticals, and functional foods, with emphasis on plant secondary metabolites and the Mediterranean diet. Author of over 230 publications, he has advanced understanding of antioxidants, essential oils, and phytotherapeutics in preclinical and clinical contexts. He serves on editorial boards; is a member of the World Academy of Sciences, Asian Council of Science Editors, and Italian Pharmacological Society; and is a founding member of the Italian Society of Environmental Medicine. A dedicated mentor, he exemplifies integration of traditional phytochemistry with modern biomedical sciences.

Contributors

Maryam Aftab Department of Biosciences, COMSATS University Islamabad, Islamabad, Pakistan

Tayyaba Afzal Department of Animal Nutrition and Feed Science, Faculty of Biology and Animal Science, Wrocław, Poland

Rina Agustina Department of Pharmacy, Faculty of Pharmacy, Hasanuddin University, Makassar, Indonesia

Nosheen Akhtar Department of Agricultural Sciences, Texas State University, San Marcos, TX, USA

Department of Botany, PMAS, Arid Agriculture University, Rawalpindi, Pakistan

Awais Ali Barari Natural Resources, Abu Dhabi, United Arab Emirates

Sidra Amin Department of Genetics, Discipline of Biological Sciences, Faculty of Biology and Animal Science, Wroclaw University of Environmental and Life Sciences, Wrocław, Poland

Umme Ammara Department of Botany, Rawalpindi Womn University, Rawalpindi, Pakistan

Muhammad Umair Asghar Department of Animal Nutrition and Feed Science, Faculty of Biology and Animal Science, Wrocław University of Environmental and Life Sciences, Wrocław, Poland

Tehreem Baig Department of Biomedical Engineering and Sciences, School of Mechanical and Manufacturing Engineering, National University of Sciences and Technology, Islamabad, Pakistan

Ayesha Batool School of Ecology and Nature Conservation, Beijing Forestry University, Beijing, China

Nurhasni Hasan Faculty of Pharmacy, Hasanuddin University, Makassar, Republic of Indonesia

Sobia Ikram South Australian Research and Development Institute, Adelaide, SA, Australia

Marcello Iriti Department of Biomedical, Surgical and Dental Sciences, University of Milan, Milan, Italy

Muhammad Azhar Jameel Department of Zoology, Wildlife, & Fisheries, PMAS-Arid Agriculture University, Rawalpindi, Pakistan

Barari Natural Resources, Abu Dhabi, United Arab Emirates

Muhammad Salman Khalid Department of Allied Health Sciences, IQRA University Chak Shahzad Campus, Islamabad, Pakistan

Muhammad Israr Khan Institute of Agriculture and Life Sciences, Gyeongsang National University, Jinju-si, South Korea

Muhammad Jawad Khan Department of Biomedical Engineering and Sciences, School of Mechanical and Manufacturing Engineering, National University of Sciences and Technology, Islamabad, Pakistan

Shahid Ullah Khan Integrative Science Center of Germplasm Creation in Western China (CHONGQING) Science City and Southwest University, College of Agronomy and Biotechnology, Southwest University, Chongqing, People's Republic of China

Gil Won Kim Institute of Agriculture and Life Sciences, Gyeongsang National University, Jinju-si, South Korea

Nimra Mudassir Department of Biomedical Sciences and Engineering, School of Mechanical and Manufacturing Engineering, National University of Science and Technology, Islamabad, Pakistan

Ambreen Najaf Department of Botany, Rawalpindi Women University, Rawalpindi, Pakistan

Jaroslaw Prockow Institute of Biology, Faculty of Biology and Animal Science, University of Environmental and Life Sciences, Wrocław, Poland

Sawaira Rahman Department of Botany, Women University Mardan, Mardan, Pakistan

Luay Rashan Biodiversity Research Centre, Dhofar University, Salalah, Oman

Iqra Riaz Department of Botany, PMAS Arid Agriculture University, Rawalpindi, Pakistan

Saboon Department of Botany, Women University Mardan, Mardan, Pakistan

Qurat ul ain Sajid Department of Animal Nutrition and Feed Science, Faculty of Biology and Animal Science, Wrocław University of Environmental and Life Sciences, Wrocław, Poland

Shifa Shaffique Department of Applied Biosciences, Kyungpook National University, Daegu, Republic of Korea

Asim Shahzad School of Industry and Urban Construction, Qingdao Hengxing University of Science and Technology, Qingdao, China

Raheem Shahzad Department of Horticulture, University of Haripur, Haripur, Pakistan

Naima Shahzadi Department of Botany, Government College Women University Sialkot, Sialkot, Pakistan

Isha Shakoor Department of Plant Protection, Wroclaw University of Environmental and Life Sciences, Wrocław, Poland

Adeeb Shehzad Biodiversity Research Centre, Dhofar University, Salalah, Oman

Sohail Department of Agriculture, Forestry, and Bioresources, College of Agriculture and Life Sciences, Seoul National University, Seoul, South Korea

Rahmat Wali Barari International Limited Company, Specialist Plant Taxonomy & Biodiversity, Section-Barari KSA-Branch, Riyadh, Kingdom of Saudi Arabia

Asim Waris Department of Biomedical Sciences and Engineering, School of Mechanical and Manufacturing Engineering, National University of Science and Technology, Islamabad, Pakistan

Iram Zaheer Department of Environmental Sciences, Fatima Jinnah Women University, Rawalpindi, Pakistan

Rahees Zaheer Department of Biosciences, COMSATS University Islamabad, Islamabad, Pakistan

Department of Allied Health Sciences, DAKSON Institute of Health Sciences, Islamabad, Pakistan

Shah Zareen School of Advanced Biotechnology, Global Plant Global Stress Research Center, Konkuk University, Seoul, South Korea

Part I

Phytochemistry and Drug Discovery

Foundation of Phytochemistry and AI-Driven Innovations in Drug Discovery

1

Nosheen Akhtar, Muhammad Azhar Jameel, Awais Ali, and Asim Shahzad

Abstract

The use of natural compounds for the treatment, prevention, and management of various diseases is as old as the origin of humans. The subsequent discoveries of bioactive compounds occurred in the nineteenth and twentieth centuries, which developed the foundation of phytochemistry. Phytochemicals are bioactive compounds with pharmacological properties. They offer health benefits and have high demand across industries, including pharmaceuticals, nutraceuticals, cosmetics, etc. The complexity of phytochemicals hinders their isolation, purification, and characterization. In addition, their low stability, lower yield, and toxicity also pose limitations in their comprehensive analysis. Traditional methods like spectrometry and chromatography have been used for the discovery of phytochemicals, but they are time-consuming and have low detectability. AI transformed phytochemistry by enabling efficient structural elucidation and metabolite profiling of compounds. They process large-scale data sets with high accuracy.

N. Akhtar (✉)
Department of Agricultural Sciences, Texas State University, San Marcos, TX, USA

Department of Botany, PMAS, Arid Agriculture University, Rawalpindi, Pakistan

M. A. Jameel
Department of Zoology, Wildlife, & Fisheries, PMAS-Arid Agriculture University, Rawalpindi, Pakistan

Barari Natural Resources, Abu Dhabi, United Arab Emirates

A. Ali
Barari Natural Resources, Abu Dhabi, United Arab Emirates

A. Shahzad
School of Industry and Urban Construction, Qingdao Hengxing University of Science and Technology, Qingdao, China

Y. Bibi et al. (eds.), *Artificial Intelligence in Phytochemistry*,
https://doi.org/10.1007/978-981-92-0875-3_1

It speeds drug discovery by screening thousands of compounds in a short duration to predict their structure and biological activities. The algorithms use supervised and unsupervised data and can identify novel compounds. This chapter provides an overview of the origin of phytochemistry and how AI solutions are transforming conventional methods into data-driven, optimized, and expeditious processes.

Keywords

Phytochemicals · Bioactive compounds · Artificial intelligence · Drug discovery · Metabolite profiling

1.1 Introduction

Plants have been used for centuries in traditional medicine due to their therapeutic properties and they are a part of cultural history. Plant-based products shaped modern pharmacology. According to an estimate, approximately 70–80% of people around the globe use traditional healing remedies and consider them safe, accessible, or affordable. Phytomedicines are prepared from natural resources and promote healing with less or no side effects. These medicines are either directly extracted from plants or structurally modified; However, many phytochemicals that impart positive effects on human health remained unidentified [1]. Plants are categorized based on the composition and structure of these chemicals. Secondary metabolites in plants have unique structures and are essential for plant growth and development. They are produced from the primary metabolites by various pathways. Phytochemistry deals with the study of chemicals derived from plants. It describes their structure, function, synthesis, and mode of action [2]. Phytochemicals have antioxidant potential to scavenge reactive oxygen species (ROS) and decrease oxidative stress. A diet full of phytochemicals (fruits, grains, vegetables, legumes, and nuts) lower the chances of cardiovascular diseases, cancer, and various other illnesses. There is a diversity of phytochemicals with their potential use in different fields [3].

1.2 Historical Overview

Natural products or bioactive compounds have unique structural properties. They are obtained from algae, fungi, bacteria, and plants, and less than 10% of sources have been evaluated. The ancient traditional medicines, including Unani, Chinese medicines, and Ayurveda in the Middle East, have ethnobotanical uses of plants for various diseases [4]. A Sumerian clay slab is the oldest documentation of drugs, and it is estimated to be 5000 years old. It contains 12 formulations of drugs by using over 250 medicinal plants [5]. The Egyptian pharmaceutical record Ebera Papyrus, which dates to 2900 B.C., includes the use of more than 700 plants to treat human illness. They utilized the plant-based formulations as ointments, infusions, pills, and gargles. In history, the earliest record in the form of a clay tablet shows that people in 2600 B.C. use oil obtained from *Commiphora* species and *Cupressus sempervirens*. The use of natural products is also documented in Chinese history. Their well-known text includes Shennong Herbal, ~100 B.C., with information on 365 drugs, Materia Medica, 1100 B.C., containing details of 52 descriptions, and Tang Herbal, 659 A.D., which describes 850 drugs. Dioscorides, 100 A.D., was a Greek physician who documented the collection, uses, and storage of herbal medicines. The historical knowledge about the medicinal properties of bioactive compounds is based on trial-and-error methods. Humans have been exploring available materials for centuries to treat diseases. They also do palatability tests to evaluate the properties of bioactive compounds. *Salvia*, a native American plant, was used by Indian tribes during childbirth. It has been documented that sweet gum from the leaves and stems of *Alhagi maurorum* consists of sucrose, inert sugar, and melezitose. It is sugary in taste and oozes from the stems and leaves of plants at high temperatures. It is documented that this plant is used to treat leprosy, epistaxis, obesity, constipation, and anorexia. In Israel, people used the root decoction for bloody diarrheas, while Roman and Konkani people used that plant for nasal polyps and asthma. *Ligusticum scoticum* was eaten raw to increase immunity against infections [6].

Plant therapeutic properties have been explored over millennia, which led to the development of the current healthcare system. The timeline of phytochemical discovery has been given in Fig. 1.1. The isolation of tartaric acid from grapes laid the foundation of phytochemistry in 1769. In history, one of the most powerful plants was *Papaver somniferum*, and investigators were curious to analyze the effects of morphine [7]. Humans started extracting active phytochemicals from plants at the start of the nineteenth century. The finding of quinine from *Cinchona* is one of the

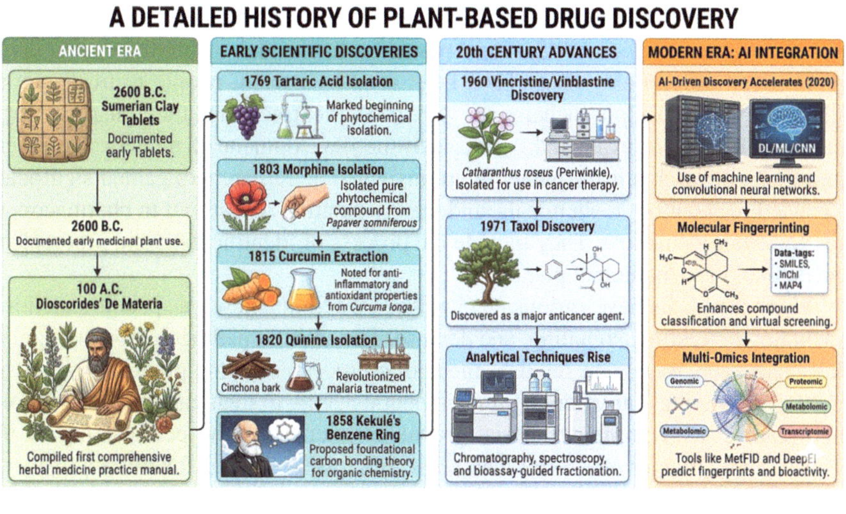

Fig. 1.1 Timeline of phytochemicals discovery

remarkable discoverys. Researchers were passionate about exploring new medicines for human healing. Laboratories were created to extract the active compounds from medicinal plants and transferred to pharmaceutical companies. The drugs discovered before World War II include hyoscine and atropine from Solanaceae, digitoxin from leaves of *Digitalis purpurea* (used for cardiac arrhythmias and heart failure), codeine and morphine from *Opium* poppy (1803). Besides, curcumin from *Curcuma longa* (1815) demonstrated antioxidant and anti-inflammatory properties and was also used to treat Alzheimer's. The source of colchicine (1819) is *Colchicum autumnale* and is used to treat inflammation. Additionally, acetylsalicylic acid is derived from the bark of *Salix* spp. and has antipyretic, analgesic, and anti-inflammatory properties. Vincristine and vinblastine (1858) were obtained from *Catharanthus roseus* and used to treat different types of cancer, such as leukemia, Hodgkin's disease, and lymphoma [2, 4]. In World War II, first antibiotic *Penicillium* (1928) was developed to treat bacterial infection. It was a game-changer and saved countless lives from bacterial infections. Different drugs were isolated from various species of *Penicillium* [8].

1.3 Origin of Phytochemistry

The knowledge of medicinal plants has been transferred among indigenous communities and from one generation to another, which forms the base of Phytochemistry. Phytochemistry has been established as a separate discipline to focus on natural compounds found in plants, algae, fungi, and marine life. It includes research activities involving plant biomolecules or chemicals derived from plants, especially secondary metabolites. It deals with properties, structural conformation, and their mode of action in organisms. The researchers focus on the isolation and testing of novel metabolites with potential industrial use in pharmaceuticals, alternative medicines, traditional medicines, dietary supplements, and nutraceuticals. Many drug discoveries have origins in traditional remedies. It is evident from history that many plant-based modern-day drugs have been used by ancient people to treat various illnesses. In the past, phytochemicals were used as extracts, and their pharmacological properties were evaluated, while with the development of modern equipment and tools, phytochemistry gained more insight. The researchers gained significant knowledge about phytochemicals, and the information was stored in pharmacopoeias and Materia medica. They include the process of identification and protocols to prepare and use medicinal plants [2].

Phytochemicals have various mechanisms of action and serve specific functions, including protection, signaling, and defense. They inhibit pathogen growth, regulating gene expression, and regulate metabolic processes. Some phytochemicals are used as chemopreventive or chemotherapeutic agents to treat cancer. Essential oils and plant extracts inhibit bacterial growth by increasing the permeability of the phospholipid bilayer of the cell membrane and loss of cellular functioning [9].

1.4 Major Categories of Phytochemicals

Plants produce a wide variety of phytochemicals, and thousands of phytochemicals have been discovered and used for various purposes. Some of the important phytochemicals are carotenoids, saponins, isoprenoids, polyphenols, phytosterols, dietary fibers, and polysaccharides. Phytochemicals have been categorized into three broad groups: terpenoids, alkaloids, and phenolics (Fig. 1.2). Terpenoids contain a structurally diverse range of metabolites that are obtained from repetitive fusion of isoprene units (C_5H_8). Their biological activities make them an important candidate in pharmaceuticals. The famous terpenoid-containing drugs include artemisinin (antimalarial) and Taxol (anti-cancer), obtained from *Artemisia annua* and *Taxus brevifolia*, respectively. Some terpenoids have been used as fragrances and flavors in cosmetics and fruits. Many living organisms synthesize terpenes to perform various physiological functions. Their role includes plant growth, development, defense, and reproduction. Gibberellin is a diterpene and plant hormone that regulates seed germination, plant stature, and flowering. Alkaloids are heterocyclic nitrogen-containing compounds obtained from amino acids. They are important natural compounds with a range of medicinal properties, including antiasthma, pain reliever, antibacterial, analgesic, and antiarrhythmic. The word "alkaloid" was given because its properties are similar to those of inorganic alkalis. It is the largest group of nitrogen-bearing secondary metabolites with over 21,000 compounds. They exist as a mixture of some major and various minor alkaloids. Examples include vincristine and vinblastine obtained from *Catharanthus roseus* and morphine from *Papaver somniferum*. They are bitter in taste and contain a variety of molecular structures. Phenolics contain a hydroxyl group (–OH) that binds with a diversity of compounds (makes complexes with metals, proteins, and other molecules). This reactivity supports their binding with free radicals and contributes to defense against oxidative stress. They also have anti-inflammatory and anti-microbial properties. The main groups of phenolics include phenylpropanoids, tannins, lignin, and anthocyanin [10–12].

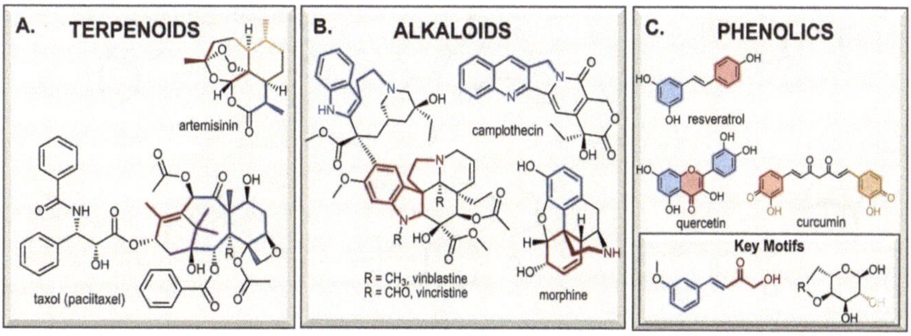

Fig. 1.2 Main groups of phytochemicals

1.5 Techniques Used for Drug Discovery

Medicinal plants have active ingredients that play a significant role in religious rituals, traditional medicines, and gastronomy. In 1858, August Kekule presented the linking of carbon atoms to develop chains and cyclic structures. This discovery shifted the attention of the investigator to the structure of active plant chemicals. Many new questions originated, but few were answered. One of the important questions was whether there is any relation between the structure and properties of active plant compounds. Some phytochemicals were easily analyzed for structure, while complex compounds took years. For example, strychnine was discovered in 1818, and its structure elucidation took 127 years of work. It contains a high number of nitrogen atoms, which makes it complex [7]. In the current era, consumers are inclined toward plant-based chemicals not only for treating diseases but also as food additives. The plant-based ingredients have a unique structure that imparts good effects on health. These active compounds can be used as food and feed additives, and they decrease the chances of bacterial resistance as compared to antibiotics. They are also used as antioxidants and preservatives in the food industry. Their active functional activities improve food quality and shelf life. The demands for plant-based active ingredients or phytochemicals boost scientific research to uncover the active compounds and develop innovative products [13].

The structures of some important drugs discovered in the nineteenth and twentieth centuries are shown in Fig. 1.3. The phytochemicals have been isolated from various plant parts by using different solvents and fractions to remove the impurities. Then the phytochemicals are separated by chromatography and bioassay-guided separation. The phytochemicals are identified by using High-Pressure Liquid Chromatography (HPLC), UV-Visible Spectroscopy, Fourier Transform Infrared Spectroscopy (FTIR), Gas Chromatography-Mass Spectrometry (GC-MS), and Nuclear Magnetic Resonance (NMR) Spectroscopy. HPLC separates the

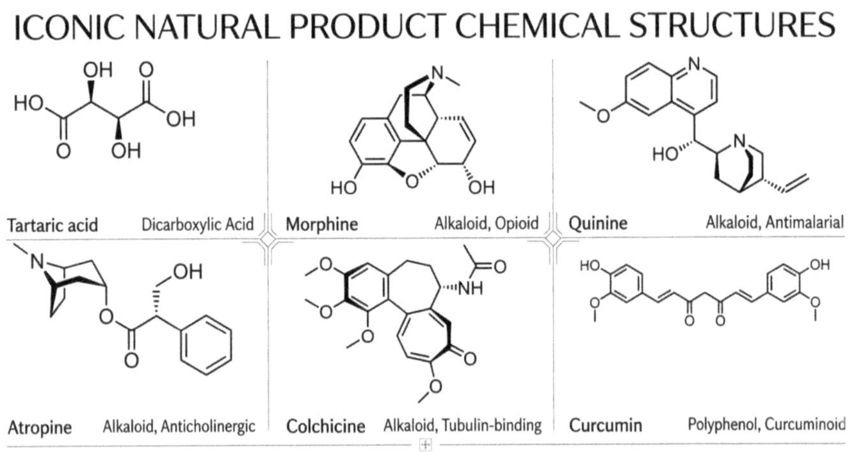

Fig. 1.3 Structure of some important phytochemicals

compounds when they interact with a stationary phase (tightly packed column of solid particles) and a mobile phase (solvent). Pressure of up to 400 bars is required to move the analyte through the column, and then it moves to the detector. It is suitable for the compounds that denature at high temperatures and cannot be vaporized. In a single operation, it performs the quantitative and qualitative evaluations of the sample. In UV-Visible spectroscopy, a beam of light passes through the solvent, and the absorbance is recorded at a specific wavelength. It characterized the absorption, transmission, and reflection of different pigments and is useful in evaluating the concentration of the analyte solution. FTIR analyzes the absorption spectra generated by photochemical reactions and identifies the functional groups of compounds. It screens the samples quantitatively and qualitatively. Samples can be analyzed in solid, liquid, and gaseous states through GC-MS, and it gives qualitative and quantitative data of phytochemicals. NMR shows the physical, chemical, and biological characteristics of compounds. Usually, one-dimensional techniques have been used by chemists to analyze the chemical structure; complex structures require two-dimensional techniques [14]. These methods are time-consuming and labor-intensive. For instance, a cancer drug obtained from the yaw tree, Taxol, took 30 years. One of the reasons is the lower quantity of these bioactive compounds in plants and their structural complexity, which makes the discovery complex. They show instability, low solubility, and toxicity, which further make the trials challenging for researchers. They need reference data, and this makes the identification of novel compounds difficult. Nowadays, testing of samples for phytochemicals has been done on a large scale, but it is costly and slow and produces results with low accuracy. The equipment identifies the drugs based on desired properties or functional groups. They do not give information about the behaviors of compounds in the bodies of humans or other living organisms [15].

1.6 AI and Drug Discovery in the Modern Era

Advancements in science and technology facilitate the integration of AI in drug discovery. AI uses various technologies, such as image processing, machine learning, and deep learning, to keep pace with the discovery of phytochemicals. They identify the structure, molecular properties, and map the genotype to phenotype ratio that affects their biosynthetic pathways. The integration of Computer-Assisted Structure Elucidation (CASE) and multi-omics enhanced the capacity to model phytochemicals and to understand the role of genetic variation in chemical diversity. AI has been used in virtual screening to analyze the biological properties of phytochemicals and to evaluate the binding of molecules at the target site. They also predict molecules with high binding potential. In silicon studies, these techniques facilitate analysis of the antimicrobial, anti-inflammatory, and antioxidant properties of phytochemicals [16]. AI also includes genome and metabolomic mining and predict the functional and structural properties of drugs. Omics data reveal the genetic pattern and pathway of phytochemical biosynthesis, such as transporters, resistance genes, and their link with primary metabolism, whereas nuclear magnetic resonance and metabolomics can explore the chemical features of metabolites. Metabolomics can identify the biosynthetic component, while genome mining algorithms can show the biosynthetic potential, so in combination, this system works to explore the genetic basis (gene or metabolite structure) and biosynthesis of phytochemicals (Fig. 1.4) [17, 18].

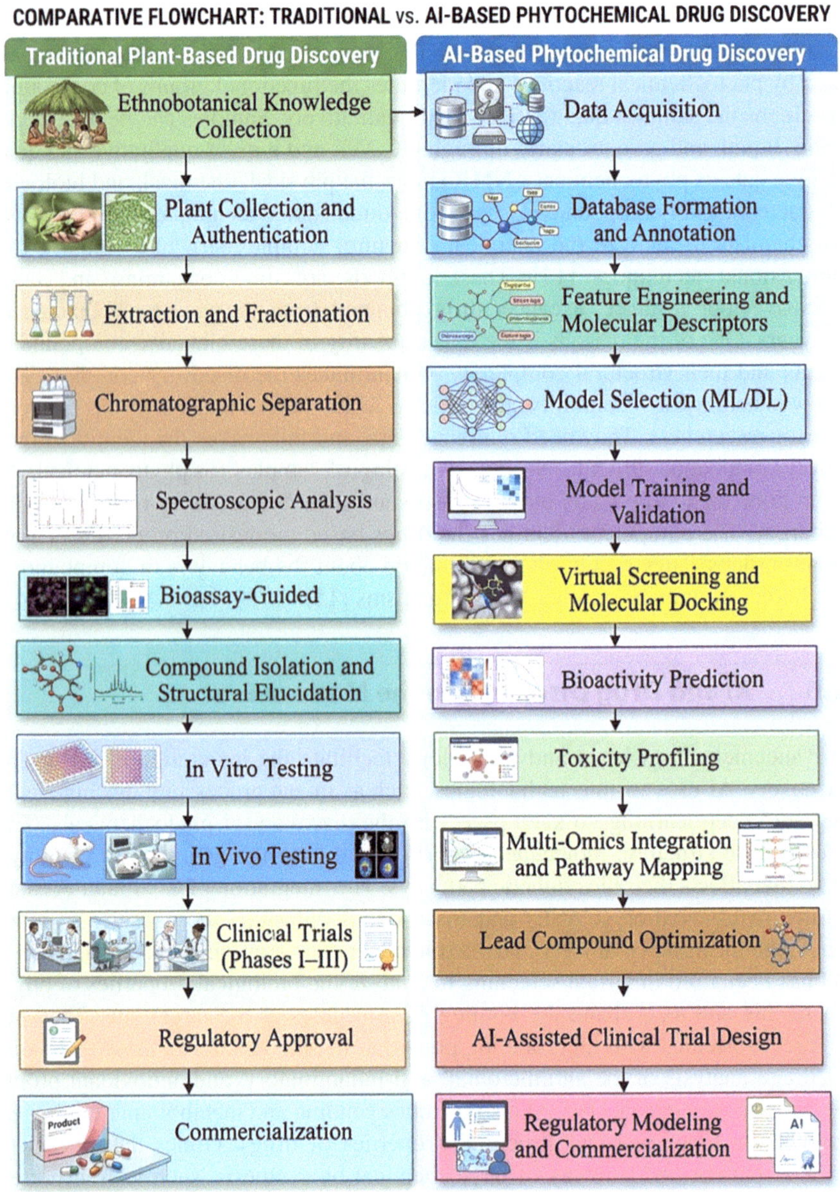

Fig. 1.4 Traditional vs AI-based drug discovery workflow

1.6.1 Characterization of Molecular Structure

The identification of the molecular structure of phytochemicals is done by spectral technology. It analyzes the spectral information of unknown compounds and identifies correlations of descriptors, including the number of peaks, maximum absorption of peaks, spacing among peaks, and frequency range, with the existing data. The use of neural networks in AI is a suitable approach to analyze complex spectral data. It involves three steps. Initially, they predict the spectra and discover the main compounds through a database. In the second step, based on spectral data, the researchers foresee the structure of molecules, while in the third step, AI uses computational models and algorithms and develops the structure of molecules, which can result in the identification of novel compounds [19].

1.6.2 Machine Learning

AI transforms drug discovery by efficiently processing data. Machine learning (ML) is a powerful tool that can recognize the biologically complex patterns of bioactive molecules. In computational methods of drug discovery, ML and novel algorithms speed up the process of identification and optimization of desired drugs. They also improve the capability of molecular docking, virtual screening, and quantitative structure-activity relationships (QSAR) [20]. ML uses trained and untrained datasets that facilitate the discovery of novel compounds. ML uses programming languages, and researchers can easily perform this function in Python and R. ML techniques that have been used in drug discovery include K-means, principal component analysis (PCA), artificial neural networks (ANN), support vector machines (SVM), random forest, decision tree, and k-Nearest Neighbor (kNN). These techniques are simple and computationally demanding and enhance the prediction accuracy in comparison with traditional predictive algorithms [21]. AI has been used at various stages of drug discovery, including identification, drug-drug interaction, and optimization of clinical trials. Deep learning and ML are useful in predicting the toxicology of drugs, their design, and safety profiling. Generative models and CNN are prominent in the prediction of protein structure and drug design. Reinforcement learning and transfer learning reduce the time and cost of drug discovery by predicting molecular properties and optimizing drugs [22]. MetFID is a tool in metabolomics that can take the data of MS/MS spectra and evaluate their patterns. It analyze the data of known compounds, how they break apart, and what their molecular structure is. It can take the estimated fingerprints to train ANN. So, the estimated fingerprints from the database can be used to predict the unknown compounds. DeepEI is a computer-based program that identifies unknown compounds based on their ionization-mass spectra. This program is trained on database spectra of chemical compounds and connects the mass spectra of chemical compounds with their molecular structure. They rank the molecules based on their similarity of molecular fingerprints with the data entries [19].

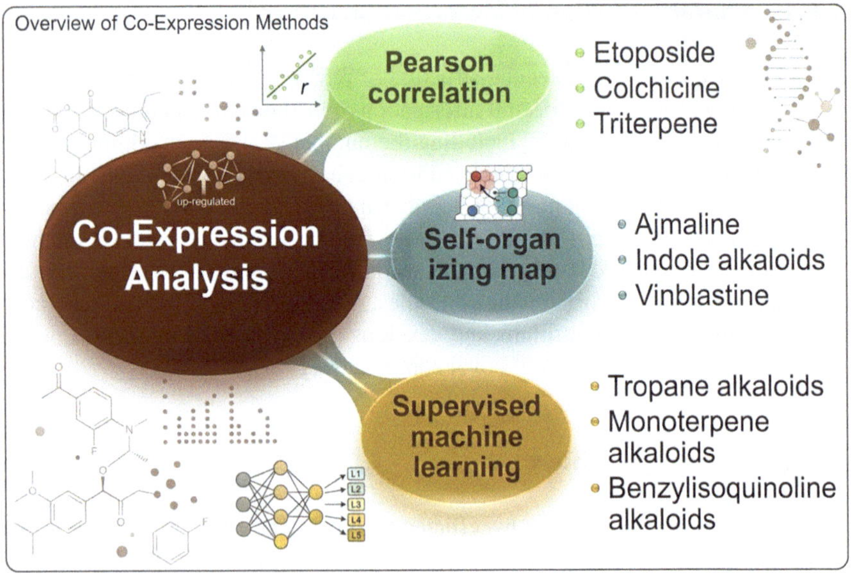

Fig. 1.5 Multi-omics tools for analyzing phytochemicals

The metabolic pathways of plant metabolites have been elucidated by using different tools, as shown in Fig. 1.5. Data mining techniques and machine learning have been utilized for expression-based identification, homology-based identification, and genomic proximity identification. Correlation analysis is important for the discovery of pathways of natural compounds. It utilizes statistical tools to pinpoint the gene that gives out similar expression patterns. It is assumed that the enzymes and genes that follow the same pathway show similar expression patterns. Pearson correlation was effectively used to untangle metabolic genes involved in the synthesis of different metabolites. The unsupervised ML algorithms, like Self-organizing maps (SOM) based on expression patterns, group the data points and facilitate the discovery of genes. It has been used to identify cytochrome P450 and 2 isomerases that take part in the biosynthesis of quassinoid and limonoid from *Ailanthus altissima*, and another P450 involved in the biosynthesis of xanthone and benzophenone. When no information is available, supervised ML algorithms can be useful to make predictions about unknown steps of pathways. Feed-forward neural networks and artificial neural networks are used to identify the genes that take part in the biosynthesis and transportation of compounds. The algorithms of these methods were trained on expression patterns of already known biosynthetic genes [23].

The chemical and structural features of molecules are identified by molecular characterization and uploaded to a computer in a machine-readable format. Initially, molecular characterization techniques, including SMiles ARbitrary Target Specification (SMARTS), International Chemical Identifier (InChl), and Simplified Molecular Input Line Entry System (SMILES), are developed to store and retrieve molecular data and align the common molecular characteristics among different molecules. The similarity of the target molecule has been determined either by ligand molecules or through receptor proteins. The advanced methods have been created to gain insight into molecular structures. Molecular fingerprinting shows that the molecules, pharmacophore-based descriptors, and circular fingerprinting patterns identify the similarities among shapes. The latter points out structurally related molecules, while the former facilitates scaffold hopping and virtual screening. MAP4 fingerprint patterns integrate the idea of substructure and atomic pairs and are useful in distinguishing phytochemicals or plant-based products from bacteria and fungi. The biological activities of molecules have been identified by using a quantitative structure-activity relationship in k-merize' SD shaped technique [17].

1.6.3 AI in Action: Modelling for Phytochemicals

To extend the rate of discovery of phytochemicals and broaden their application, it is crucial to utilize advanced machine learning techniques to predict the structure and activities of phytochemicals. Kim and Kwon [24] designed a framework based on machine learning to identify and characterize phytochemicals. It contains two independent models, the first of which was a neural network designed to predict the bioactivities of phytochemicals that fall into four major categories, that is, lipid metabolism, neurotoxicity, anti-inflammatory activity, and antioxidant activity. The second model is a CNN that processes the proteins for amino acid sequences and predicts the Enzyme Commission (EC) number, which is linked with the function of the enzyme. The bioactive predictions of the first model show an accuracy of 97%, while the latter model achieves a prediction efficiency of 81%. The data collected from this technique can be utilized in various fields of drug development. Lee et al. [25] designed a machine learning-based model that predicts the biological activity of phytochemicals. This model includes a fully connected neural network, which can efficiently analyze complex nonlinear patterns of molecules. The dataset has different categories of biological activity of molecules. The model performance was analyzed based on four metrics: Precision, Recall, F1-Score, and Accuracy. The best performance was given by the model with a depth of four layers. The model presents the highest true positive ratio for Lipid metabolism, while toxicity has the lowest true negative ratio. This network achieved an F1-Score of 0.72 and an accuracy of 0.79 on the public dataset.

Essential oils are organic in nature (terpenes and terpenoids) and extracted from plants. They are used in pharmaceuticals, cosmetics, food, and beverages. They are also part of aromatherapy to enhance relaxation and health conditions. El-Attar et al. (2020) conducted a study to predict the biological activities of essential oil-containing plants by using Multiclass Neural Network (MNN) and Convolutional Neural Network (CNN). MNN is a machine learning technique that classifies the input into different categories, while CNN is a deep learning technique that extracts the features from grid-like data such as images. They implement these two algorithms and develop a biological activity prediction model. The testing shows an accuracy of 81% and 98% for MNN and CNN, respectively. The CNN was reliable due to its high accuracy, so it was adopted to further evaluate the bioactivities of essential oil-containing plants. The final prediction shows an overall accuracy of 97% which supports the use of deep learning for predicting the bioactivities of plants.

The molecular and chemical properties of natural compounds can be used to predict their medicinal value. Yoo et al. [26] propose a deep learning approach in which they use molecular and chemical characteristics of natural compounds, and in addition, they also take latent information from literature to fill the incomplete knowledge. They apply a partially connected deep neural network approach to capture and analyze the complex and heterogeneous information in multiple hidden layers of the model. The input data included molecular features, chemical properties, and latent knowledge, and the output was the medicinal effect of drugs. The deep learning model generated or predicted the medicinal uses of 4507 compounds for 15 diseases. This model can be used in the initial screening of compounds from a bunch of natural compounds on a large scale.

Phytochemicals have antioxidant potential that plays an important role in reducing oxidative stress. The traditional methods to identify antioxidant compounds are time-consuming and labor-intensive. Jung et al. [27] used only the molecular structure of compounds in a ML model to analyze the antioxidant potential of molecules. The assessment was done for five ML algorithms: Logistic Regression (LR), Deep Neural Network (DNN), Support Vector Machine (SVM), Random Forest (RF), and XGBoost. The data set contains more than 1900 compounds with antioxidant activities. RF and SVM show high accuracy and effectively separate active and inactive compounds. This study confirms that ML can expedite the discovery of phytochemicals with antioxidant properties that lead to the development of effective drugs.

1.7 Challenges and Suggestions

There are several challenges that need to be addressed to get maximum benefit from AI. AI models use large and accurate databases; however, biased sampling and missing data can affect model prediction and result in skewed predictions. So, it is crucial to train the models on high-quality data sets with a diversity of plant samples. The AI models that identify plants and their medicinal uses (due to active phytochemicals or ingredients) are trained in controlled conditions, and their output or results accuracy in a natural setting can be reduced due to variability in the environment. This problem can be resolved by generating a context-rich database that can control or nullify the effects of environmental variables. The models should interpret and explain the results with high accuracy [16]. Limited data is available to train the AI models, as the databases usually lack comprehensive datasets on biological activities, structures, and their therapeutic properties. The models face data scarcity, which needs to be addressed. Phytochemicals are involved in multiple mechanisms and have different targets, so AI can struggle to analyze this complexity and result in low prediction efficiency. Human advancement in collecting the relevant information by experimentation will refine ML models in the future [15].

1.8 Conclusion

The discovery of novel phytochemicals is important for creating effective therapeutic drugs. The complexity of phytochemical compounds due to intricate structure or multiple functional groups makes the identification challenging for researchers, but AI uses machine learning and computational approaches like molecular fingerprinting to assist in the prediction of structure and biological activities. The learning and computational potential of AI enhances the discovery of phytochemicals. It can save time, energy, and process large data sets with high accuracy. This chapter summarizes the progress made by AI to uncover the structure and function of plant-based active metabolites. The research gap and limited data availability decrease the quality of the database, which in turn affects the reliability and accuracy of AI models. The researchers from different fields, including pharmacology, biochemistry, agricultural sciences, food sciences, and plant sciences, should work in an interdisciplinary manner to bridge the gap between existing knowledge and modern technology. They can create new strategies for AI modern applications. Interdisciplinary research can play a significant role in the generation of an accurate and diverse database for training AI models. Therefore, AI can transform the industries that use phytochemicals and provide better services to the public.

References

1. Chihomvu, P., et al. (2024). Phytochemicals in drug discovery—A confluence of tradition and innovation. *International Journal of Molecular Sciences, 25*(16), 8792.
2. Sinha, D., et al. (2023). Phytochemistry history and progress in drug discovery. In *Phytochemistry computational tools and databases in drug discovery* (pp. 1–26). Elsevier.
3. Pawase, P. A., et al. (2024). A conceptual review on classification extraction bioactive potential and role of phytochemicals in human health. *Future Foods, 9*(1), 1–13.
4. Marzouk, M. M., et al. (2025). Historical overview of pharmacognosy and phytochemistry. In *Pharmacognosy and phytochemistry: principles, techniques, and clinical applications* (pp. 1–16). Wiley.
5. Sinha, P., & Yadav, A. K. (2025). Unraveling the anti-breast cancer activity of Cimicifugae rhizoma using biological network pathways and molecular dynamics simulation. *Molecular Diversity, 29*(1), 241–254.
6. Dias, C., et al. (2015). Phytochemistry and activity against digestive pathogens of grape (Vitis vinifera L.) stem's (poly) phenolic extracts. *LWT-Food Science and Technology, 61*(1), 25–32.
7. Cordell, G. A. (2011). Phytochemistry and traditional medicine–a revolution in process. *Phytochemistry Letters, 4*(4), 391–398.
8. Phillipson, J. D. (2001). Phytochemistry and medicinal plants. *Phytochemistry, 56*(3), 237–243.
9. Doughari, J. H. (2012). *Phytochemicals: Extraction methods, basic structures and mode of action as potential chemotherapeutic agents*. INTECH Open Access Publisher.
10. Campos-Vega, R., & Oomah, B. D. (2013). Chemistry and classification of phytochemicals. In *Handbook of plant food phytochemicals* (pp. 5–48). Wiley-Blackwell.
11. Eljounaidi, K., & Lichman, B. R. (2020). Nature's chemists: The discovery and engineering of phytochemical biosynthesis. *Frontiers in Chemistry, 8*, 596479.
12. Koche, D., Shirsat, R., & Kawale, M. (2016). An overerview of major classes of phytochemicals: Their types and role in disease prevention. *Hislopia Journal, 9*(1/2), 1–11.

13. Socaci, S. A., Fărcaş, A. C., & Tofană, M. (2020). Functional ingredients derived from aromatic plants. In *Feed additives* (pp. 133–146). Elsevier.
14. Banu, K. S., & Cathrine, L. (2015). General techniques involved in phytochemical analysis. *International Journal of Advanced Research in Chemical Science, 2*(4), 25–32.
15. Gangwal, A., & Lavecchia, A. (2025). Artificial intelligence in natural product drug discovery: Current applications and future perspectives. *Journal of Medicinal Chemistry, 68*(4), 3948–3969.
16. Saripalli, H. R., et al. *Journal of Population Therapeutics & Clinical Pharmacology.*
17. Duan, Z.-Y., et al. (2024). Moslae Herba: Botany, traditional uses, phytochemistry, and pharmacology. *Molecules, 29*(8), 1716.
18. Blanco-Gonzalez, A., et al. (2023). The role of AI in drug discovery: Challenges, opportunities, and strategies. *Pharmaceuticals, 16*(6), 891.
19. Varghese, R., et al. (2025). Artificial intelligence driven approaches in phytochemical research: Trends and prospects. *Phytochemistry Reviews*, 1–16.
20. Barbudhe, V., et al. (2025). Artificial intelligence-based crop recommendation, and disease detection system. *International Journal, 9*(11).
21. Elbadawi, M., et al. (2024). The role of artificial intelligence in generating original scientific research. *International Journal of Pharmaceutics, 652*, 123741.
22. Barbudhe, V. K., Zanjat, S. N., & Karmore, B. S. (2025). Transforming global health outcomes through smart technology and the SGD framework. In *Driving global health and sustainable development goals with smart technology* (pp. 473–490). IGI Global Scientific Publishing.
23. McConnachie, M., et al. (2025). From data to discovery: Leveraging big data in plant natural products biosynthesis research. *The Plant Journal, 122*(6), e70288.
24. Kim, K., & Kwon, K. (2025). A systematic review of the evaluation in K-12 artificial intelligence education from 2013 to 2022. *Interactive Learning Environments, 33*(1), 103–131.
25. Lee, H., et al. (2025). Recent advances in AI-based toxicity prediction for drug discovery. *Frontiers in Chemistry, 13*, 1632046.
26. Yoo, J., et al. (2023). Industrializing AI/ML during the end-to-end drug discovery process. *Current Opinion in Structural Biology, 79*, 102528.
27. Jung, W., et al. (2024). Absorption distribution metabolism excretion and toxicity property prediction utilizing a pre-trained natural language processing model and its applications in early-stage drug development. *Pharmaceuticals, 17*(3), 382.

Part II

AI Simulation Techniques in Phytochemistry

AI-Powered Prediction of Molecular Properties of Phytochemicals

2

Iqra Riaz, Shahid Ullah Khan, and Sohail

Abstract

For thousands of years, nature remained a source of medical substances, as plants have long been used as medicine to cure different health issues, while their phytochemicals have inspired the design, discovery, and development of new drugs but also present formidable obstacles for computational and experimental assessment. Conventional analyses remain expensive, time-consuming, and poorly scalable for large natural product libraries, despite the importance of accurately predicting key molecular properties, such as solubility, lipophilicity, permeability, bioavailability, and toxicity, in evaluating their therapeutic relevance. The emergence of artificial intelligence (AI), especially machine learning and deep learning, has enabled effective and data-driven modeling of phytochemical properties using both representation-learning techniques and descriptor-based regression frameworks. These include transformer-based chemical language models that extract long-range structural dependencies from molecular strings as well as graph and message passing neural networks that simulate molecular structures as atom–bond graphs. This chapter focuses on important AI approaches and platforms like DeepChem and Chemprop, with a view of multitask learning, transfer learning, and rigorous validation techniques, to guarantee generalization across

I. Riaz
Department of Botany, PMAS Arid Agriculture University, Rawalpindi, Pakistan

S. U. Khan (✉)
Integrative Science Center of Germplasm Creation in Western China (CHONGQING) Science City and Southwest University, College of Agronomy and Biotechnology, Southwest University, Chongqing, People's Republic of China
e-mail: shahidbiochem@wmc.edu.pk

Sohail
Department of Agriculture, Forestry, and Bioresources, College of Agriculture and Life Sciences, Seoul National University, Seoul, South Korea

Y. Bibi et al. (eds.), *Artificial Intelligence in Phytochemistry*,
https://doi.org/10.1007/978-981-92-0875-3_2

various phytochemical scaffolds. Challenges related to data scarcity, interpretability, and distributional bias are discussed alongside emerging solutions, including federated learning, foundation models, and next-generation ADMET platforms. This chapter aims to highlight a step forward toward mechanistically informed and scalable prediction frameworks, integrating AI with multi-omics and systems biology, for accelerating phytochemical-driven drug discovery.

Keywords

Phytochemicals · Artificial intelligence (AI) · Machine learning · Drug discovery

2.1 Introduction

Phytochemicals are secondary metabolites derived from plants, and they represent one of the most structurally diverse classes of compounds. Traditional medicine and contemporary drug discovery have been supported by major groups like alkaloids, flavonoids, terpenoids, saponins, and polyphenols [1]. They are appealing candidates for lead identification and optimization due to their wide range of bioactivities that include antioxidant and anticancer effects. However, drug development faces significant analytical and computational challenges due to the intrinsic structural complexities of phytochemicals, which are characterized by multiple chiral centers, aromatic systems, and glycosidic linkages [2].

An important stage in the drug development process is predicting important molecular characteristics, such as permeability, bioavailability, aqueous solubility, lipophilicity (logP), and toxicity (ADMET: absorption, distribution, metabolism, excretion, and toxicity). These characteristics have a direct linkage with pharmacokinetics and pharmacodynamics that determine whether a substance can reduce side effects while reaching its desired target site in vivo at a therapeutically related concentration [3]. Tens of thousands of structurally diverse molecules can be found in natural product libraries, making it nonlinear, expensive, time-consuming, and frequently impractical to evaluate these properties experimentally.

Cheminformatics has been revolutionized by recent developments in artificial intelligence (AI), specifically machine learning (ML), and deep learning (DL). AI-driven models can identify patterns in existing datasets with the help of high-quantity screening and expedited lead discovery [4], and hierarchical structure–property relationships can be learnt by AI methods, and they improve generalization capabilities compared to classical quantitative structure–activity relationship (QSAR) models, which rely on linear correlations between molecular descriptors and bioactivity [5]. AI models can efficiently capture plant-derived chemical features and enhance the prediction of ADMET properties by pretraining on large, diverse chemical repositories like ChEMBL, PubChem, and BindingDB, and then fine-tuning on phytochemical-specific datasets [6]. This transfer-learning approach accelerates natural product-based drug discovery, which makes it possible to quickly and virtually screen thousands of natural compounds, prioritize high-value candidates, and rationally design derivatives with enhanced pharmacological profiles.

2.2 AI Methodologies for Property Prediction

2.2.1 Neural Networks and Deep Learning

Deep learning (DL) is an algorithm that automatically learns hierarchical representations from raw data. As pointed out by Gilmer et al. [5] and Yang et al. [6], DL architectures such as Graph Neural Networks (GNNs), Message Passing Neural Networks (MPNNs), Convolutional Neural Networks (CNNs), and Recurrent Neural Networks (RNNs) have demonstrated strong performance in predicting physicochemical and ADMET properties (Fig. 2.1).

Traditional molecular representations, such as descriptors or molecular fingerprints, distill structural information into fixed number vectors. However, these manually defined features may not capture essential structural details of phytochemicals. Deep learning overcomes this limitation by learning directly from molecular graphs, where edges represent bonds and nodes represent atoms [7]. In this context, MPNNs obtain the topological and chemical context of the molecule by updating the representation for each atom iteratively using the "messages" received from surrounding atoms and bonds [5].

Chemprop is an excellent example of this approach that is a flagship platform developed by MIT and the Broad Institute. In this approach, molecular graphs are created internally by its Message Passing Neural Network (MPNN) model, after accepting SMILES strings (i.e., text representations of chemical compounds). Message passing involves the iterative exchange and aggregation of bond features, such as bond type and conjugation, and atom-level features, such as atomic number,

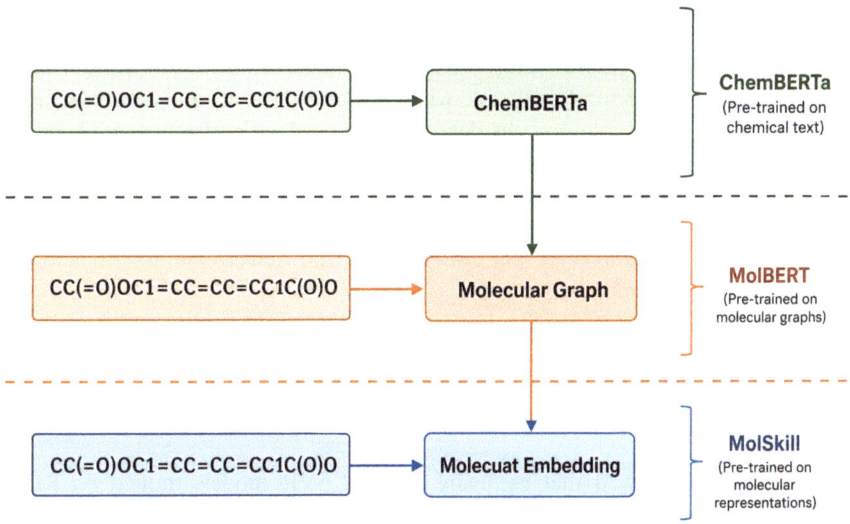

Fig. 2.1 Transformer-based chemical language model

valence, and aromaticity, throughout the molecular graph. To predict properties from the learned molecular embedding, a feed-forward neural network is used, which is a continuous vector representation from which both local chemical environments and global structural information are encoded [8]. Owing to its architecture, Chemprop is capable of predicting significant properties with utmost accuracy, which includes aqueous solubility, logP, and blood-brain barrier (BBB) permeability.

Chemprop v2.2.1 (2025) is a recent advancement in which enhanced GPU device allocation and introduced fine-tuning of foundation models—large pre-trained MPNNs trained on millions of molecules—to enable efficient large-scale screening are utilized [9]. While deep learning methods, including Chemprop, can benefit from scaffolds that were not visible for training, these models have unique capabilities for phytochemicals. And while MPNNs have become increasingly influential in molecules modeling, transformers, which are related to advancements in natural language processing, have also entered the spotlight. For example, MolBERT and ChemBERTa use self-attention mechanisms to model the context-dependent relationships that exist between atoms and structural features when encoding SMILES strings as a type of chemical "language" [10]. Such models are especially useful for natural products, which often contain unusual structural motifs that are challenging for conventional descriptors to capture.

2.2.2 Regression Models

Traditional machine learning algorithms are still essential, especially for small- to medium-sized datasets that are frequently used in phytochemical research, even though deep learning models provide more flexibility. Regression and classification tasks frequently employ models like k-Nearest Neighbor (kNN), Random Forests (RF), Support Vector Regression (SVR), and Gradient Boosting Machines (GBM) [4, 11]. They are particularly valuable when working on structurally diverse natural products because they require less data to achieve robust performance. Moreover, they are less likely to overfitting on small datasets and offer interpretable insights into feature-property relationships.

Regression models are generally dependent on molecular descriptors that provide numerical values to represent a single molecule's electronic, topological, and physicochemical characteristics.

Programs like RDKit, PaDEL-Descriptor, or Mordred are utilized to compute descriptor sets. These sets include parameters like Topological Polar Surface Area (TPSA), logP (lipophilicity), electronic properties (e.g., HOMO–LUMO gap), molecular weight (MW), number of hydrogen bond donors/acceptors, and rotatable bond count.

Studies have showed that by using RF and SVR models trained on RDKit descriptors, accurate prediction of solubility and partition coefficient (logP) for several phytochemicals can be done [12]. Because Random Forest models are resistant to overfitting and offer feature importance scores that indicate which descriptors contribute most to property predictions, they are especially helpful for determining the most significant molecular features. When it comes to capturing nonlinear structure property relationships, Gradient Boosting Machines (GBMs), which use repeated decision trees to lower prediction error, frequently perform better than linear models. Despite being less flexible than deep learning methods, descriptor-based models provide interpretability, which is essential for comprehending predictions for complex natural products. As a result, hybrid methods that combine descriptor-based regression with deep feature extraction like graph embeddings are becoming more popular [13].

2.3 Real-Time Tools and Platforms

Open-access infrastructures with pre-established virtual screening, molecular representation, and model training tools have provided substantial support to AI-driven cheminformatics. Two of the more prominent frameworks for predicting molecular properties for synthetic and natural compounds are DeepChem and Chemprop.

2.3.1 DeepChem

DeepChem is an open-source Python package aimed at enabling deep learning in materials science and chemistry [14]. It not only provides methods for featurization, handling datasets, and evaluation but also incorporates multiple machine learning architectures, such as feed-forward networks, graph convolutions, and multitask networks. Multitask learning is especially useful for phytochemicals because it enables shared chemical information across related molecular properties to enhance predictions for compounds with limited experimental data. This method allows single model to predict various properties at once by using correlation between molecular properties.

This feature is particularly useful in investigating phytochemicals, where labeled data is often insufficient. Ramsundar et al. [14] presented multitask models that predict relevant endpoints such as toxicity, logP, and solubility, to provide predictive accuracy and allow information transfer across tasks. Another quality of DeepChem is transfer learning, whereby the model is pre-trained on large data sets and fine-tuned on a smaller domain specific phytochemical dataset.

A representative study used DeepChem's GraphConvModel to predict the aqueous solubility of flavonoids from *Camellia sinensis*, or green tea. With an RMSE of 0.42 on an external test set, this deep graph convolutional network proved that deep learning can function well on comparatively small datasets [15]. DeepChem's adaptable architecture offers a complete platform for molecular feature extraction, property prediction, and virtual screening in natural product discovery.

2.3.2 Chemprop

Chemprop employs a scalable MPNN architecture with both regression and classification capabilities to predict molecular properties. Because of its Python API and simple command-line interface, it may be used for high-throughput applications (Fig. 2.2).

Significant improvements were added in the 2025 update (v2.2.1), including fine-tuning foundation models (allows Large pretrained molecular embeddings), better GPU/CPU device allocation (speeds up large compound libraries), and better quantification of uncertainty (crucial for evaluating the accuracy of model predictions).

One of the real world application of Chemprop's is forecasting the hepatotoxicity of alkaloids isolated from *Berberis vulgaris*. This model outperformed traditional descriptor-based models with an AUC of 0.89 using curated data from Tox21 and BindingDB [16, 17]. This example shows how well deep learning captures minute structural patterns linked to toxicity.

Additionally, new tools are becoming more and more useful for phytochemical analysis such as ADMET lab 2.0, ChemBERTa, and MolSkill. These models combine deep neural representations with conventional QSAR interpretability [18].

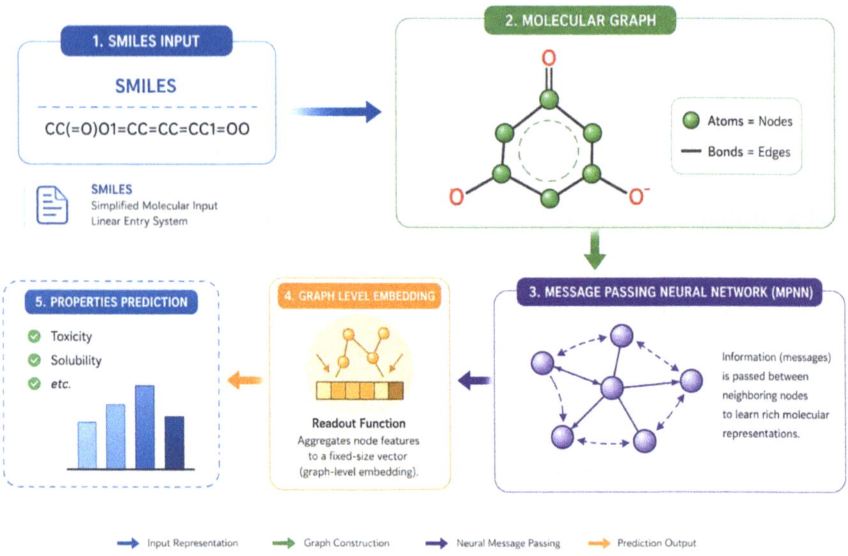

Fig. 2.2 Chem-prop architecture

2.4 Model Validation and Performance Metrics

AI models require rigorous testing to assure that they can handle new data sets, other than the data they were trained on. A common issue that is faced during working with small data sets of phytochemicals is overfitting, as they memorize limited set of data instead of learning generalized pattern, thus reducing their usefulness.

2.4.1 Cross-Validation Techniques

K-fold cross-validation is commonly used method that is used for evaluation of model performance by dividing the dataset into *k* non-overlapping subsets (folds). In each iteration, *k–1* folds are utilized for model training, whereas the other fold is booked for testing. The same procedure is repeated *k* times to assure that every molecule is used exactly once for validation [3]. Typically, five- or ten-fold cross-validation is employed, balancing computational efficiency with robust performance estimation. By averaging results across all folds, this approach reduces variance arising from random data partitioning and provides a more reliable assessment of model generalization. *However, for structurally diverse natural products, scaffold-based k-fold splitting is increasingly recommended*, as it prevents closely related molecular scaffolds from appearing in both training and test sets, thereby offering a more realistic evaluation of a model's ability to generalize to novel chemical spaces.

2.4.2 Performance Metrics

Metrics specific to the prediction task are used to assess the model's performance. While classification tasks usually depend on accuracy, RMSE and R^2 are frequently used for regression. However, some molecular classes are underrepresented in phytochemical datasets, which are frequently small and unbalanced. In these situations, the Area Under the Precision–Recall Curve (AUCPR) better captures model performance for rare or minority classes, making it especially useful for natural product studies, and the Matthews Correlation Coefficient (MCC) offers a balanced measure of prediction quality across all classes.

Metrics that vary depending on the prediction task are used to assess model performance.

2.4.2.1 Regression Tasks

In regression tasks, several key metrics are used to evaluate model performance. Root Mean Square Error (RMSE) is used to measure deviation between experimental and the predicted values. The Mean Absolute Error (MAE) shows the average magnitude of errors and R^2 (Coefficient of Determination shows what proportion of the variance can be explained by the model.

2.4.2.2 Classification Tasks

In classification tasks, one of the most widely used evaluation metrics is the Receiver Operating Characteristic–Area Under the Curve (ROC-AUC), which assesses the model's capacity for discrimination [19].

2.4.3 External Validation and Benchmarking

In order to make sure the generalization of the model, separate datasets are used for external validation. Frequently used repositories for phytochemical research include Natural Product Activity and Species Source Database (NPASS) that includes activity data for more than 35,000 natural product entries [20], SuperNatural II offering the expected pharmacological activity of natural products along with their annotated structures, COCONUT (Collection of Open Natural Products) that is an open-access resource collecting various natural products from diverse sources, and LOTUS that is centered on natural product data that is open, FAIR (Findable, Accessible, Interoperable, and Reusable) [2].

Cross-domain validation supports accurate predictions for novel phytochemicals by ensuring that models are not over-specialized to particular compound classes.

2.5 Challenges and Future Directions

The use of AI in phytochemical research still confronts significant obstacles, despite substantial improvements.

2.5.1 Data Scarcity and Imbalance

Many phytochemicals have sparsely labeled datasets due to the lack of comprehensive ADMET and bioactivity profiles [21]. Supervised learning is made more difficult by the predominance of some property classes (like "non-toxic") in data imbalance. Semi-supervised learning approaches and data augmentation strategies, such as SMILES augmentation, which creates different SMILES representations for the same molecule to increase the dataset, are used more frequently to improve model performance and generalizability in order to address these problems.

2.5.2 Bias in Training Data

The prevalence of synthetic compounds in the majority of public datasets, including ChEMBL, PubChem, and Tox21, leads to distributional bias. Models trained mainly on synthetic chemo types, such as benzene derivatives, amides, and heteroaryl scaffolds, often show reduced precision when applied to natural products, which frequently contain unique scaffolds like flavonoid backbones, alkaloid heterocycles, or terpenoid ring systems [4]. Because of this scaffold bias, predictions are less accurate for compounds that are not part of the training set's dominant frameworks, such as complex polycyclic or macrocyclic natural products.

2.5.3 Emerging AI Solutions

2.5.3.1 Transfer Learning

Transfer learning mitigates the lack of data by fine-tuning on phytochemical data and pretraining on large synthetic datasets. This approach produces accurate predictions even in the absence of labeled natural product data by utilizing general chemical information found in the model's weights [14].

2.5.3.2 Federated Learning

Federated learning (FL) shields intellectual property and spheres data privacy by enabling multiple institutions to work together to train models without sharing raw data. Each institution trains a local model on its own phytochemical dataset in a typical FL workflow, and only the model updates—such as weights or gradients—are sent to a central server. These updates are combined by the server to create a global model, which is then sent to participants for additional local training. While maintaining the decentralization of sensitive data, this iterative process produces a potent shared model (Fig. 2.3).

Emerging foundation models like ChemGPT and MegaMolBART can be incorporated by federated learning. These models, which have been pretrained on extensive chemical corpora, offer rich molecular representations that can be refined

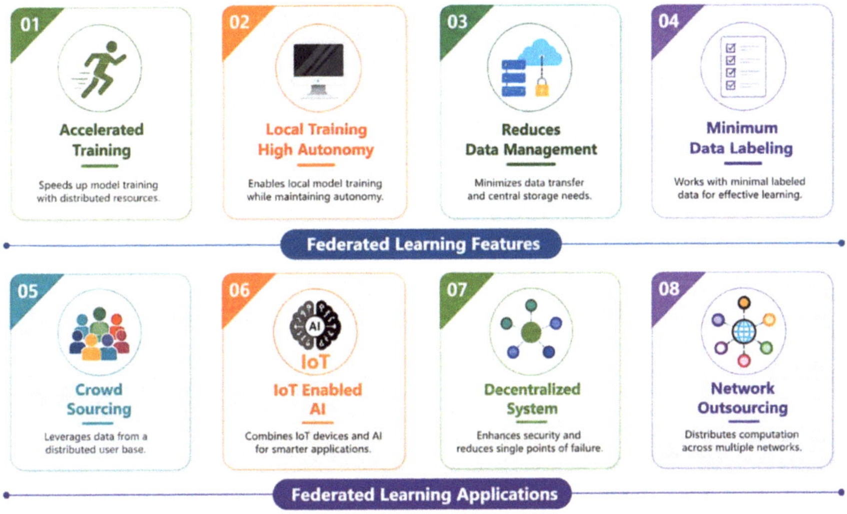

Fig. 2.3 Federated learning in phytochemical research

across dispersed phytochemical datasets, improving ADMET and property prediction without sacrificing data privacy.

2.5.4 ADMET-AI Platform

In 2024, Stanford developed ADMET-AI, a next-generation machine learning system for large-scale pharmacokinetic and toxicity prediction [18]. It evaluates chemical libraries, including natural products, by combining explainable AI modules, multitask transformers, and graph neural networks. The explainable AI modules support regulatory compliance and decision-making by offering comprehensible insights into which molecular features underlie particular predictions (Fig. 2.4). Regulatory approval in drug discovery processes depends on the interpretability made possible by ADMET-AI's modular design.

2.5.5 Integration of Multi-omics and Systems Biology

In future, projects with AI models will be used to combine large data sets, which can predict molecular properties along with biological activities and mechanisms of action [18]. By linking phytochemical characteristics to target protein interactions and pathway disruptions, researchers can build thorough predictive models to detect the therapeutic properties of natural products. Similarly, metabolite gene

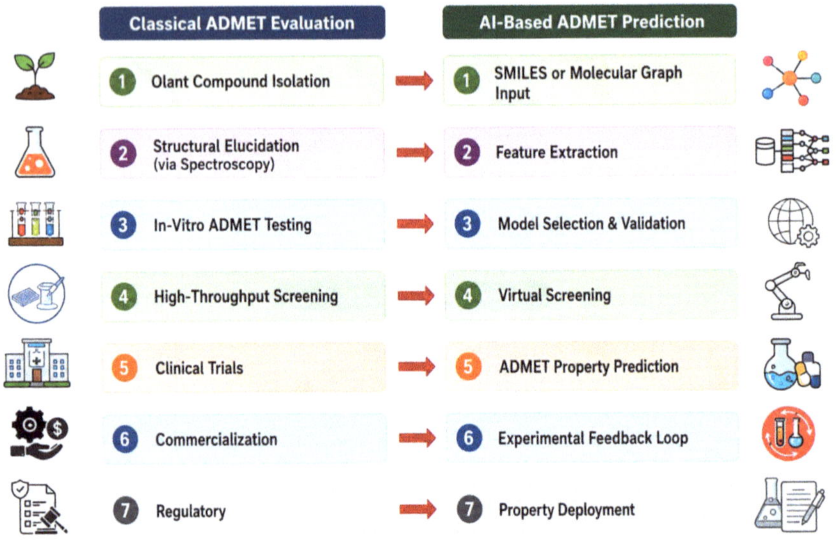

Fig. 2.4 Traditional vs AI-based ADMET prediction

co-expression networks can indicate the effect of a particular phytochemical on metabolic and regulatory pathways by improving mechanistic understanding and directing drug discovery.

2.6 Summary

Deep learning architectures such as message passing neural networks (MPNNs), graph transformers, and hybrid ML models allow researchers to explore hidden chemical insight by using both small and big data. These developments are transforming the way of predicting molecular properties for phytochemicals and connecting the worlds of computational drug discovery and classical natural product chemistry. However, to truly harness AI in phytochemical research, the challenges of data poverty, bias, and explanation ability must be tackled. The growing convergence of AI, federated data infrastructures, and multi-omics integration heralds an era of intelligent phytochemical discovery.

References

1. Bolzani, V. D. S., et al. (2012). Natural products from Brazilian biodiversity as a source of new models for medicinal chemistry. *Pure and Applied Chemistry, 84*(9), 1837–1846.
2. Rutz, A., et al. (2022). The LOTUS initiative for open knowledge management in natural products research. *Elife, 11*, e70780.
3. Tropsha, A., et al. (2024). Integrating QSAR modelling and deep learning in drug discovery: The emergence of deep QSAR. *Nature Reviews Drug Discovery, 23*(2), 141–155.
4. Coley, C. W., et al. (2019). A graph-convolutional neural network model for the prediction of chemical reactivity. *Chemical Science, 10*(2), 370–377.
5. Gilmer, J., et al. (2017). Neural message passing for quantum chemistry. In *International conference on machine learning*. PMLR.
6. Yang, H., et al. (2019). admetSAR 2.0: Web-service for prediction and optimization of chemical ADMET properties. *Bioinformatics, 35*(6), 1067–1069.
7. Ahmed, E., et al. (2017). Secondary metabolites and their multidimensional prospective in plant life. *Journal of Pharmacognosy and Phytochemistry, 6*(2), 205–214.
8. Yang, B., et al. (2019). Condconv: Conditionally parameterized convolutions for efficient inference. *Advances in Neural Information Processing Systems, 32*. https://proceedings.neurips.cc/paper/2019/hash/f2201f5191c4e92cc5af043eebfd0946-Abstract.html
9. Graff, D. E., et al. (2026). Chemprop v2: An efficient, modular machine learning package for chemical property prediction. *Journal of Chemical Information and Modeling, 66*(1), 28–33.
10. Schottlender, G., et al. (2025). Beyond Tanimoto: A learned bioactivity similarity index enhances ligand discovery. *Frontiers in Bioinformatics, 5*, 1695353.
11. Tian, L., et al. (2024). Investigating the asymmetric impact of artificial intelligence on renewable energy under climate policy uncertainty. *Energy Economics, 137*, 107809.
12. Tian, P., et al. (2022). Estimation model of global ionospheric irregularities: An artificial intelligence approach. *Space Weather, 20*(9), e2022SW003160.
13. Zhou, L., et al. (2023). Predictable artificial intelligence. *arXiv preprint*. https://arxiv.org/abs/2310.06167
14. Ramsundar, B., et al. (2019). *Deep learning for the life sciences: applying deep learning to genomics, microscopy, drug discovery, and more*. O'Reilly Media.

15. Sen, C. K. (2025). Artificial intelligence tools in biomedical research: Part 1-literature search and knowledge mining. *Antioxidants & Redox Signaling, 44*, 1.
16. Ahmed, H. M. (2025). Enhancing image classification using graph attention networks. *Iraqi Journal for Computers and Informatics, 51*(2), 33–45.
17. Dutta, A., & Barupal, D. (2025). Predicting toxicity and bioactivity of the chemical exposome: A case study for the blood exposome database. *bioRxiv*.
18. Zhao, Y., He, X., & Wan, Q. (2024). Combined machine learning models, docking analysis, ADMET studies and molecular dynamics simulations for the design of novel FAK inhibitors against glioblastoma. *BMC Chemistry, 18*(1), 203.
19. Shi, H., et al. (2023). Improving software defect prediction in noisy imbalanced datasets. *Applied Sciences, 13*(18), 10466.
20. Zeng, X., et al. (2018). NPASS: Natural product activity and species source database for natural product research, discovery and tool development. *Nucleic Acids Research, 46*(D1), D1217–D1222.
21. Zhou, Y., et al. (2024). Obacunone, a promising phytochemical triterpenoid: Research progress on its pharmacological activity and mechanism. *Molecules, 29*(8), 1791.

Simulation of Drug-Phytochemical Interactions Using Artificial Intelligence (AI)

3

Rahees Zaheer, Maryam Aftab, Muhammad Israr Khan, and Shah Zareen

Abstract

The prevalence for a number of diseases has developed a scenario where conventional treatment methods fail to achieve complete remedies. Researchers are exploring computer-aided drug design for therapeutic development, as developed exponentially during COVID-19. Artificial intelligence (AI) has opened new avenues for screening phytochemicals from medicinal plants based on their physicochemical properties to predict their absorption, distribution, metabolism, excretion, and toxicity (ADMET). These advancements have reduced the cost and time required to evaluate the specific targets of phytochemicals through molecular docking (MD) analysis. Computer-aided drug design has revolutionized the drug discovery process, achieving a high level of accuracy. This chapter introduces the role of AI in evaluating phytochemicals for target prediction through molecular docking, as well as the role of other computational tools involved.

R. Zaheer
Department of Biosciences, COMSATS University Islamabad, Islamabad, Pakistan

Department of Allied Health Sciences, DAKSON Institute of Health Sciences, Islamabad, Pakistan

M. Aftab
Department of Biosciences, COMSATS University Islamabad, Islamabad, Pakistan

M. I. Khan (✉)
Institute of Agriculture and Life Sciences, Gyeongsang National University, Jinju, South Korea
e-mail: israrkhan@gnu.ac.kr

S. Zareen
School of Advanced Biotechnology, Global Plant Global Stress Research Center, Konkuk University, Seoul, South Korea

Y. Bibi et al. (eds.), *Artificial Intelligence in Phytochemistry*,
https://doi.org/10.1007/978-981-92-0875-3_3

Keywords

Medicinal plants · Physicochemical properties · ADMET · Molecular docking · Target prediction

3.1 Introduction

Phytochemicals are low molecular weight (LMW) substances that contain a wide range of bioactive molecules made by plants that have always fascinated people. These natural compounds, commonly present in fruits, vegetables, herbs, and spices, have been employed for centuries in traditional medicine systems globally [1]. There are a few rough estimates and at least 250,000 plant species. Phytochemicals and their metabolites found in the root, leaf, flower, stem, and bark of plants have several pharmacological effects on humans. The exact estimation of phytochemicals is uncertain; more than 100,000 different phytochemicals have been identified. As scientific inquiry investigates the molecular pathways that facilitate the therapeutic efficacy of phytochemical substances, an increasing corpus of evidence substantiates their contribution to enhancing human health and mitigating chronic disorders [2]. The average person consumes about 1.5 g of phytochemicals per day; however vegans and vegetarians may consume significant secondary metabolites. However, there are no standardized intake guidelines for phytochemicals [3].

Phytochemicals, which are bio active molecules found naturally in plants, have an important role in health promotion, illness prevention, and effective treatment. Primary metabolites and secondary metabolites are the two broad classes of phytochemicals. Essential for their survival, development, and reproduction, primary metabolites are compounds that all living creatures make. These metabolites (including proteins and carbohydrates) play an essential role in respiration and photosynthesis, two fundamental metabolic processes. Amino acids, found in these, are fundamental for many biological functions and serve as the basis of proteins. Examples of carbohydrates include glucose, fructose, and starch, which provide both energy and structural functions to cellular compartments. Cholesterol, phospholipids, and triglycerides play a central role in the building of membranes and storage of energy. Vitamins are essential micronutrients required in small amounts as a normal part of the metabolism and this includes vitamin B12, C, and D. Organic acids such as acetic acids, citric acid, and lactic acid are involved in numerous metabolic processes. These are the most basic metabolites and are the essentials of the healthy processes of all living organisms [4].

Secondary metabolites play a central role in plant defense against biotic and abiotic stressors, such as pathogens, herbivores, and other environmental challenges despite their indirect influence on the growth and developmental processes. Moreover, they also contribute to the organoleptic characteristics of plants, which give them unique tastes, aromas, and color. Secondary metabolites are further subdivided into several classes according to their production pathways and chemical composition. There are several primary classes, including phenolics, alkaloids, terpenoids, flavonoids, and others. They play a key role in the development of novel pharmaceuticals. On top of that, they are key ingredients in traditional herbal remedies. Some life-saving and health-improving drugs are derived from these secondary metabolites. Some examples of alkaloids include the powerful painkiller morphine, the well-known antimalarial drug quinine, and the substantial anticancer drug vincristine. Terpenoids like artemisinin, a highly effective antimalarial drug, and paclitaxel, a cancer chemotherapy drug, are examples [2].

Medicinal plants have seen an unprecedented rise in popularity as a form of complementary and alternative medicine around the world. Between 10% and 50% of the population in wealthy countries uses herbal remedies on a daily basis, according to the World Health Organization (WHO) [5]. According to [6], the global herbal medicines industry is projected to expand from USD 151.91 billion in 2021 to USD 347.50 billion in 2029, a compound annual growth rate (CAGR) of 11.16%. Medicines derived from plants or their metabolites have been more commercialized due to the pharmaceutical industry's attempts to supply these compounds in pharmaceutical dosage forms. Scientific research has shown that medicinal plants are effective in treating illnesses without adverse effects, and their use has increased substantially on a worldwide scale. In comparison to synthetic active pharmaceutical ingredients (API), plant-derived phytochemicals do not possess side effects. However, the chemical components of herbal medications and supplements may interact with other drugs to provide the synergistic or inhibitory effect [7, 8]. Therefore individuals, exposed to polypharmacy regimens, should be aware of drug-drug, herbal-herbal, herbal food, or herbal-drug interactions.

The biological activity of a phytochemical is related to its chemical structure. Their interactions with biological targets and pharmacological effects are influenced by important structural factors such as stereochemistry, ring systems, and functional groups. Phytochemicals can be combined chemically with aromatic rings in order to enhance their stability and reactivity [1]. Due to the involvement of rings in redox and interactions with biological targets, it may influence its redox characteristics. The elements other than carbon including oxygen, nitrogen, and/or sulfur and by direct linkage of functional groups to the ring affect polarity and chemical nature of ring [9].

Alternate double-single bonds (chirality) in a molecule also affect the biological activity of molecules. The pharmacophore composed of spatial elements determines the interaction of molecules with their desired targets that can produce a biological response. The pharmacophore evaluation helps the scientists to evaluate the action of newly designed medicines [10]. Pharmacophores having hydrogen bond donors or acceptor are found in phytochemicals. The biological targets interact with pharmacophores by forming hydrogen bonds via hydroxyl group. Moreover, the carbonyl group of pharmacophore act as hydrogen bond acceptor and form hydrogen bonds with amino group of protein. Hydrogen bonds, π–π stacking interactions, aromatic rings, and hydrophobic interactions enhance the binding affinities of phytochemicals to their target proteins. The lipophilic character of molecules can be changed by adding methyl groups that ultimately affect absorption and metabolism [2].

For structure-activity relationship (SAR), computational and AI tools have advanced the research field, as they aid in the development and process of drug research. Docking scores evaluate the binding of phytochemicals with target protein. Quantum mechanics allows to evaluate the binding affinity and reactivity of molecules through the calculation of their electronic properties. The computational tools will help scientist to accelerate the process of finding new drug candidates and advance the research on exploring therapeutic potential of phytochemicals. AI has merged the computational chemistry to the field of pharmaceutical chemistry, and pharmaceutical drug development has been advanced. On a large scale, complex metabolomes can be analyzed to interpret their biological data to evaluate their medicinal value. Not only can we discover new plant-based medicines faster but can also discover phytochemicals more precisely and efficiently. This chapter gives an in-depth introduction to how computational tools are transforming the aspect of phytochemical research using AI. This underscores the important functions that digital technologies are playing in assisting scientists to locate, study, and maximize natural medicinal treasures.

3.2 Machine Learning Models for Drug–Phytochemical Binding Predictions

3.2.1 Machine Learning (ML)

The ML technique has made it possible to evaluate the biological activity of phytochemicals. The ML algorithms are effective for identification of intricate patterns in datasets that predict the behavior of novel compounds (Fig. 3.1). ML approaches

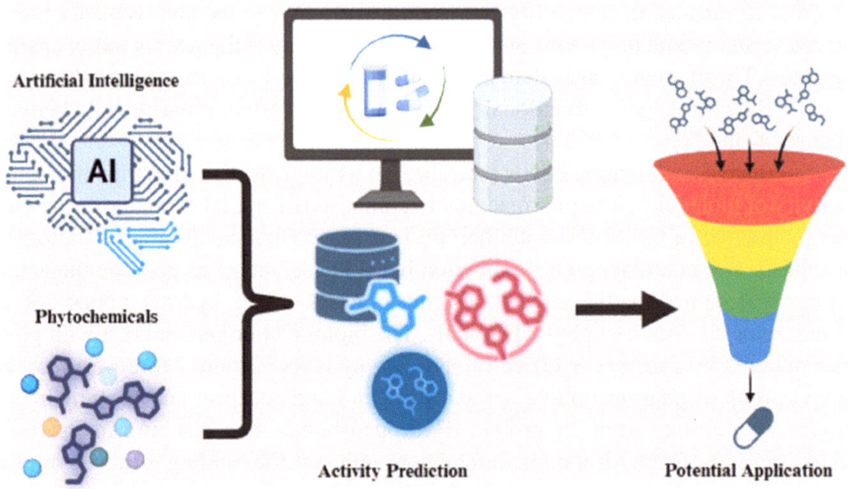

Fig. 3.1 Machine learning in phytochemical research

reveal valuable information for the development and discovery of medicinal agents. ML methods including support vector machine and neural networks facilitate to predict the phytochemical activities, pharmacokinetics, and toxicity profiles [2]. ML algorithms help in the optimization of drug development process by evaluation of databases and predicting the novel compounds biological activities.

ML involves deconstructing data, gaining knowledge from data, and then using algorithms to predict what new data sets will be like in the future. Instead of manually programming software routines with a predefined set of instructions, machines are taught to perform certain tasks by analyzing large amounts of data and using algorithms. An algorithm is written by the programmer to train the network, rather than expert rules. Algorithm performance adapts to increase in relation to the quantity and quality of training data. Consequently, ML is most effective in situations where there is a large amount of data and numerous variables, but no preexisting model or formula that links them [17].

Depending on their learning mechanism, algorithms are generally categorized as either supervised or unsupervised networks. In order to train supervised learning models, both input and output data are utilized. Using the labelled datasets continuous or categorical connections, supervised machine learning models are trained. The goal of supervised learning is to train a model that can predict future outputs given known input data. In contrast, unsupervised learning does not rely on labeled data but rather asks the model to discover patterns in the input data and utilize them to create several meaningful clusters without any human intervention [17]. By training models on many chemical components, it is feasible to construct modules that are both effective and affordable. By utilizing the comprehensive techniques of machine learning algorithms, these models reveal the essential relationships between the chemical properties of the molecule and experimental data. Following that, the experimental outcome is predicted by feeding the models the chemical data of new substances [18]. It takes a lot of patience and financial resources to use the traditional way of drug discovery. Accordingly, ML tools are quickly becoming an essential tactic for many goals in this area, such as improving biomarkers for prognosis, progression, and drug efficacy; discovering new targets; mining relevant chemical data; predicting biological traits and potentially active biological molecules; expanding our understanding of the disease; and identifying novel targets [15].

At least 80% of machine learning is devoted to data cleaning and processing, with just 20% devoted to algorithm application, according to some estimates. Consequently, for any machine learning method to be truly predictive, massive volumes of high-quality data must be readily available. Accurate, thorough, and carefully chosen training data is essential for improved predictability. It is a usual practice in experimental design to discuss the best way to estimate this parameter, including the sample size and power calculations. Whether the appropriate data is even available and what data ought to be generated experimentally are important considerations for some subjects. ML programs perform better on well-annotated, clean data that has been prepared in a systematic manner [17, 19].

3.2.1.1 Case Studies of ML-Based Models

Instead of using the conventional SAR-based approaches for predicting the activity of unknown compounds, Yue et al. [20] used a combination of gene expression data and properties of phytochemicals to construct an ML-based approach. The natural products response predictors were developed by extracting 17 natural product (NP) medicines from GDSC (Genomics of Drug Sensitivity in Cancer) across many cell lines. After screening these 17 Natural products (NPs) using 495 cell lines on average for each medication, the cell lines were categorized as either sensitive, resistant, or intermediate according to the IC50 values of the NPs.

A variety of ML models built from samples found in both resistant and sensitive groups. Randomly, 13 NPs samples were selected for 6450 cancer cell line, in which 4 NPs with 1970 cell lines were utilized as test set. Tenfold cross-validation was performed for training set that all methods revealed satisfactory results. Two antitumor NPs containing curcumin and resveratrol were selected to stop cancer cells proliferation [15, 20].

In another study, a model that predicts a database of 2892 NPs and 617 antitumor medicines to identify the NPs exhibiting anti-cancerous potential was constructed. One dimensional and two dimensional descriptors of compounds were ascertained by Molecular Operating Environment (MOE), in active and inactive datasets, respectively. The iterative stochastic elimination (ISE) method was employed for comparison of active and inactive ligands [15, 21].

3.2.1.2 Model Selection Concepts

An ML model should have the potential to generalize training data to test data. Applying an ML model's principles to data that was not used during the training phase is known as generalization. Within each strategy, there are variations in terms of the number of variables that can be managed, prediction accuracy, and training speed. Selecting algorithms carefully is essential to ensure that they are appropriate for the task at hand and the quantity and quality of data that is accessible. Considerations such as the method's signal-to-noise ratio and the amount of parameter tuning needed are also critical. When a model learns not just the signal but also some of the unusual features of the training data and incorporates them into its model, it is known as over fitting, and it has a negative effect on the model's performance on fresh data. The inability of a model to generalize to new data while simultaneously modelling the training data is referred to as under fitting.

To lessen the impact of overfitting, many people employ resampling methods or set aside some of the training data to use as validation. Ridge, LASSO, and elastic nets are regularization regression methods that punish parameters to prevent overfitting as model complexity increases. An effective method for avoiding overfitting is the dropout technique, described by Srivastava et al. [22], which involves randomly removing units from the hidden layer. Not all machine learning approaches use the same metrics to evaluate their success. Several fundamental evaluation measures can be used to compare the performance of different algorithms [23]. These include classification accuracy, kappa [24], area under the curve (AUC,) logarithmic loss, F1 score, and confusion matrix. It is highly beneficial to have access to both independently developed and gold standard data sets for building high-performing models. From desktop PCs to server clusters, a variety of hardware platforms are now accessible for high-performance mathematical calculation, with several software libraries being developed for use with central processing units (CPUs), tensor processing units (TPUs), and GPUs, among others. Some well-known ML programming frameworks are Scikit-learn, PyTorch, Keras, and the open-source framework TensorFlow, which were originally developed by engineers and researchers from Google's AI organization's Brain team [17].

3.2.2 Deep Learning in Phytochemical Research

Deep learning is a subfield of machine learning that makes use of artificial neurons to analyze data and draw conclusions. Image pattern recognition and text mining are only two of the several fields that have made use of DL (Fig. 3.2). This method is also used in drug discovery to speed up the process of drug development. Also, DL is vital in the process of developing new drugs. Lead optimization, virtual screening, ADMET features, and QSAR models are common applications [25]. Because DL techniques automate challenging processes that were previously achievable only with human methods, the field of plant identification has made tremendous strides forward. Convolutional Neural Networks (CNNs) are widely used in plant identification because they automatically extract hierarchical features from raw visual data. Using images of leaves, blossoms, fruits, and other plant features, CNNs can identify the species of plants using visual representations such as shape, texture, and color. Models such as ResNet, VGGNet, and AlexNet have demonstrated substantial improvement in accuracy when used on large plant identification data sets as compared to more traditional machine learning techniques (e.g., Decision Trees and Support Vector Machines). These designs include a plethora of convolutional filter layers, pooling procedures, and fully linked layers to get complex

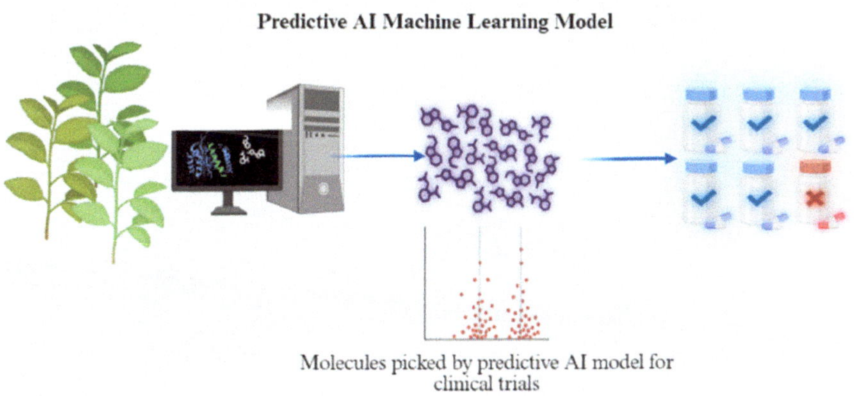

Fig. 3.2 Deep learning architecture

representations of the plant characteristics. Notable applications include transfer learning, which enables the refinement of deep learning models that have been already learnt from broad datasets of pictures, for example, ImageNet on plant specific datasets. This method helps alleviate some of that difficulty. Research has shown that utilizing plant-specific data to fine-tune models significantly improves classification accuracy, even with limited datasets [26].

Deep learning systems can forecast the potential medicinal effects of natural compounds. For both synthetic and naturally occurring chemicals, the algorithm consists of four stages, that is, collecting various types of pharmacological and natural compound data from public sources; extracting latent knowledge, molecular interactions, and chemical property features from the collected data using text mining, network analysis, and chemical property analysis; training a deep learning model with features of both approved and experimental drugs as inputs and their indication data as outputs; applying the trained model to predict the medicinal uses of natural compounds (Fig. 3.3).

To methodically build the deep learning model, a multi-step methodology was employed. Collecting and organizing experimental bioactivity data against SARS-CoV 3CLpro from PubChem Bio assays was the initial step. The next step was to turn chemical structures into descriptive terms that could capture physicochemical and structural properties. For evaluation of the model, the datasets were divided into two subsets, the training and the validation subsets. Three convolutional layers that contain max-pooling and fully connected layers make up the CNN architecture. By modifying the hyper parameters, the performance of the model was improved. By using descriptors when training the model and closely monitoring accuracy and loss

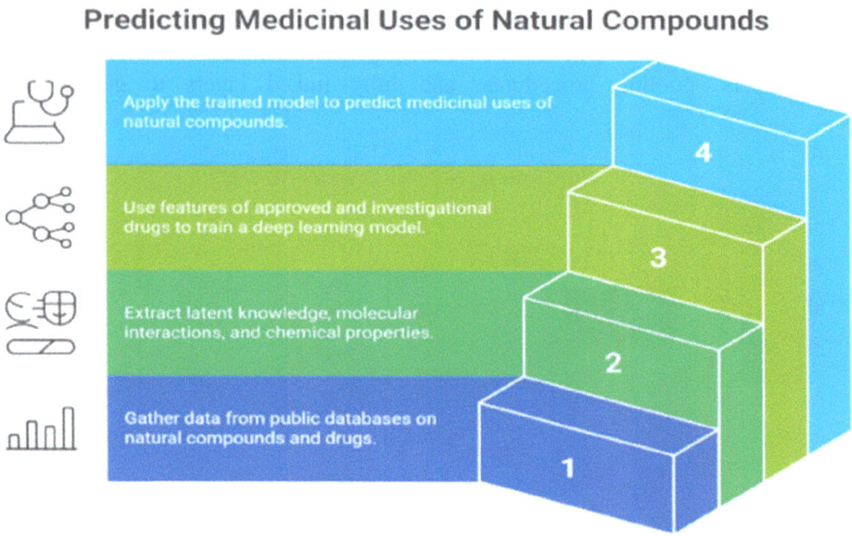

Fig. 3.3 Systematic procedure for predicting medicinal effects of natural compounds using deep learning

curves to make sure they were converging accurately, an internal accuracy of 86% on unknown data was achieved. After validation of the CNN, it was used for the virtual screening of large compound libraries composed of natural products, phytochemicals, and FDA-approved pharmaceuticals using Lipinski's rule of five to prioritize drug-like compounds. This end-to-end process showed that deep learning shows the potential to accelerate drug discovery through the creation of a powerful predictive tool that could quickly identify potential anti-SARS-CoV medications [25].

3.2.3 AI in Phytochemical Research

AI plays an important role in structure-activity relationship (SAR) studies because it enables researchers to use data-driven molecular analysis to explore molecular relationships between the structures of phytochemicals and their biological actions. Algorithms that have been trained on data from plants can be used to predict the bioactivity, toxicity, and pharmacokinetics of a medicine, which is helpful in the process of drug development. It also aids structural elucidation such as finding chemical structures, predicting their interactions with biological targets. AI also improves and accelerates the development of medication through the combination of molecular network analysis and phytochemical databases. Additionally, novel bioactive compounds that can be used in medicinal applications are also discovered when plant metabolic pathways are unraveled using AI-driven genomics and metabolomic approaches [27, 28].

The drug development process is undergoing a revolution with the use of AI-driven systems to analyze large libraries of phytochemicals and predict bioactivity and toxicity by high throughput screening. AI improves molecular docking and virtual screening, which aids the identification of possible medicinal candidates from plant compounds. Deep learning improves SAR prediction even further with the modelling of complex interactions between phytochemical substances and their biological effects. Furthermore, AI-driven metabolomics combines an enormous volume of metabolomic data and identifies new bioactive compounds and their therapeutic potential. HerbIntel and CASE are two examples of AI-powered databases that provide curated phytochemical data to help in SAR analysis that speeds up the search for natural substances that could be used as medicines [29].

Modern research depends highly on information processing technologies based on AI and sophisticated measurements for the discovery of potential bioactive chemicals and detailed information on groups of target components. There has been a significant change in the research methods with the growing usage of AI algorithm into NP chemistry. These algorithms have several uses, such as the systematic classification of NPs into different chemical and medicinal classes, the efficient detection of compounds, and the extraction of compounds [30].

The aim of AI is to program a computer to do something generally done by a human such as learning, reasoning, problem solving, perception, language interpretation, and decision making. Numerous domains make use of AI, like computer vision, robotics, automated systems, and natural language processing (NLP), which makes use of LLMs like ChatGPT. AI has a great impact on the molecular design, SAR studies, and drug development activity prediction fields. For example, activity prediction is directly influenced by AI-based categorization exercises that can rapidly identify compounds having desirable properties and then sort molecules based on properties such as toxicity or drug candidacy. Predicting continuous values, such as medication potency or binding affinity to a particular protein, is the specialty of regression tasks, and they are very important for optimizing molecular designs and for SAR research [31].

Identifying drug-target interactions (DTIs) is important in the medical and healthcare fields as drug discovery is a challenging endeavor. DTIs elucidate the processes underlying the therapeutic drug binding interactions with the target proteins and determine the binding ability of the molecule. Typical approaches to the identification of DTIs rely heavily on labor-intensive and costly in vitro and in vivo experiments. AI and other new approaches to computing, however, have made remarkable progress in this domain. Among the many benefits of AI technologies are their fast prediction rates, reduced costs, and excellent forecast accuracy [32].

3.2.3.1 Databases, Tools, and Evaluation Metrics

Data on target proteins, ligand molecules, and interactions are available in several databases, which are utilized to forecast DTIs. According to Table 3.1, these databases are often categorized as either drug-focused, target-focused, or binding databases. Both ChEMBL [35] and DrugBank [37] are notable databases that focus on drugs. The most current edition of DrugBank Online (5.1.12, published on March 14, 2024) contains 17,085 pharmaceuticals. Among these, 2786 small-molecule medications have been approved, 1629 biologists have been approved, and a plethora of other compounds are currently being studied in clinical or experimental

Table 3.1 Commonly used databases in drug–target interaction prediction

Database name	Description
Drug-focused databases	
Drug Bank	A repository for information regarding medicinal compounds and their therapeutic targets, including chemical, pharmacological, structural, and sequencing details
ChEMBL	A carefully curated collection of bioactive compounds with medicinal potential, complete with binding, functional, and other information
ZINC20	An annotated database of chemicals that are commercially available
PubChem	An extensive collection of chemical information, including structural details, characteristics, and numerous attributes
Binding interaction databases	
BindingDB	A database that contains protein binding affinities to tiny compounds that are similar to drugs
Target-focused databases	
Protein Data Bank	Proteins 3D structure database
AlphaFold Protein Structure Database	A database containing AlphaFold2-predicted protein structures

Table 3.2 Commonly used tools for physiochemical analysis

Tool name	Description
Tools for computing feature vectors	
ProtDCal	Protein features can be computed using a tool named ProtDCal using three-dimensional structures and sequences of amino acids. A total of 1998 features can be generated by ProtDCal, with 999 originating from sequence and 999 from structure
ProtParam	The ProtParam web server can determine many protein properties, including molecular weight, atomic composition, and amino acid composition, among many others
Tools for computing molecular similarities	
SIMilar COMPound (SIMCOMP)	SIMCOMP is a graph-based chemical structure comparison tool
Compound–protein interaction with R programming language (Rcpi)	Numerous structural and physicochemical characteristics of proteins and peptides might be computed by RCPI using amino acid sequences, small molecule molecular descriptors based on their topology, and descriptors of protein–protein and compound–protein interactions
Tools for generating chemical structures	
Open Babe	A number of crucial capabilities for drug development are included in Open Babel, including the ability to generate 2D and 3D molecular structures using Simplified Molecular Input Line Entry System (SMILES) strings
RDKit	One key function of the cheminformatics toolkit RDKit is the ability to create 2D and 3D molecular structures using SMILES strings. Several programming languages, such as Python and C++, are supported by RDKit

settings. The 2D structures, different properties, and bioactivities of around 2.4 million bioactive drug-like compounds are kept in ChEMBL.

Several tools are available for computing the physicochemical features of both drugs and targets. Table 3.2 presents the commonly used tools for physiochemical analysis divided into three categories.

Several criteria, each tailored to specific classification and regression tasks, are employed to evaluate the efficacy of AI-powered DTI prediction algorithms. It is a common practice that the following metrics are used to evaluate classification tasks:

1. *Positives and negatives*: Based on positive and negative labels, there would be true positives (TP), true negatives (TN), false positives (FP) and false negatives (FN), which are accurate and inaccurate predictions.
2. *Accuracy*: The percentage of accurate predictions during the entire data set is called accuracy.
3. *Precision*: Precision is a metric that calculates the ratio of correctly anticipated positive outcomes to the overall amount of anticipated positive outcomes.
4. *Receiver operating characteristic (ROC) curve*: The roc curve is a plot of the true positive rate against false positive rate at different levels of discrimination.

5. *Mean squared logarithmic error (MSLE)*: MSLE is used to measure squared logarithmic error between the predicted and actual values.
6. *Mean absolute percentage error (MAPE)*: MAPE is a measure of the accuracy of the predictions of a regression model that is not sensitive to scaling of the target variable.

3.3 Molecular Docking Simulations with AI Enhancements

Molecular docking provides valuable information to predict the binding affinities and orientations for binding of phytochemicals to their target proteins (Fig. 3.2). The receptor protein structure, obtained through NMR or X-ray crystallography, binds with the ligand, which acts as the docking molecule. Different confirmations help to simulate the phytochemical movement in binding site of proteins [13].

During COVID-19 pandemic, molecular docking has helped the scientist to evaluate the interaction and binding of phytochemicals with target proteins responsible for COVID-19 pathogenesis. In total, 35 phytochemicals and 43 medicinal agents were screened to inhibit the spike glycoproteins of coronavirus that helped to find out the compounds exhibiting the desired action. Drug probability and toxicity prediction test were passed by these compounds (Fig. 3.4). Apigenin, a phytochemical and liquid it in, a reclaimed medicine was found to be the best candidates based on ADME analysis, docking score, and drug similarity profile. Some of the

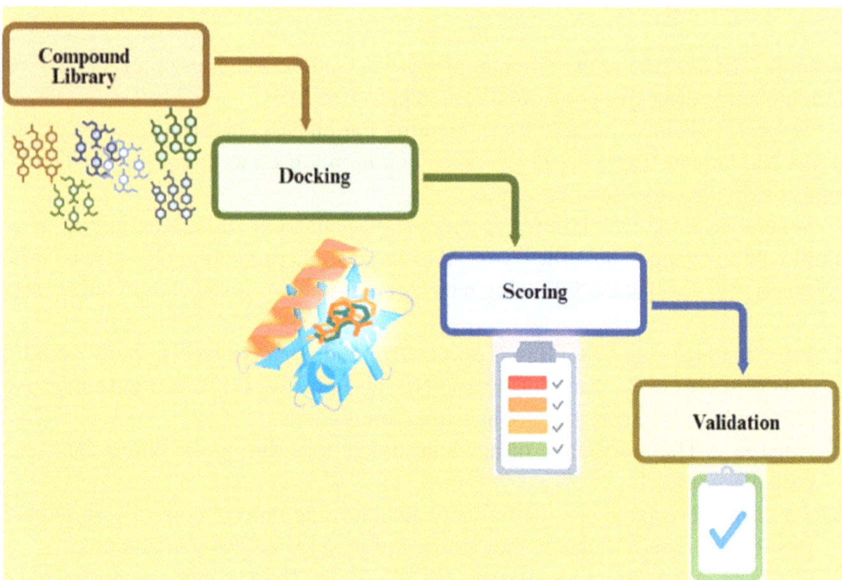

Fig. 3.4 Workflow of virtual screening and molecular docking

phytochemicals such as terpenoids, luteolin, quercetin, and phenolics have shown potential for leading. These chemicals may have the ability to block selective protein targets of SARS-CoV-2. It was proven that the way they interacted might disrupt viral pathogenesis and replication [14].

3.3.1 Molecular Dynamics (MD)

Biomolecules in the human body are dynamic, unlike the static conformations employed in classic structure-based drug discovery procedures (Fig. 3.5). As a result, molecular dynamics may be crucial for improving docking or virtual screening outcomes. Hence, it may be necessary to understand how protein molecular architectures change. Studies in drug development rely heavily on MD because it predicts structural and molecular changes in biomolecules caused by intramolecular and intermolecular interactions [15]. For understanding the interplay between phytochemicals (ligands) and the proteins that they bind to, MD simulations show how the stability and structure of the complex undergo dynamic changes. Unlike static docking techniques, which only present a snapshot of the interaction, MD simulations reproduce the movement of atoms and molecules over a preset time-period, frequently between nanoseconds and microseconds. A time-dependent study like this one lets scientists watch how solvent effects, heat mobility, and other dynamic factors cause the protein and phytochemical to change, reorganize, or conform [13].

For MD simulation, there is no shortage of software options; nonetheless, computational scientists often turn to GROMACS 2024.2, AMBER 2024, and NAMD 3.0 because of the power and accuracy of these programs in handling complex and large molecular systems. By MD simulation studies, flavonoids and G-quadruplex DNA have been focused in several studies. For cancer treatment, binding of flavonoids to G-quadruplex has been anticipated by MD simulations [16].

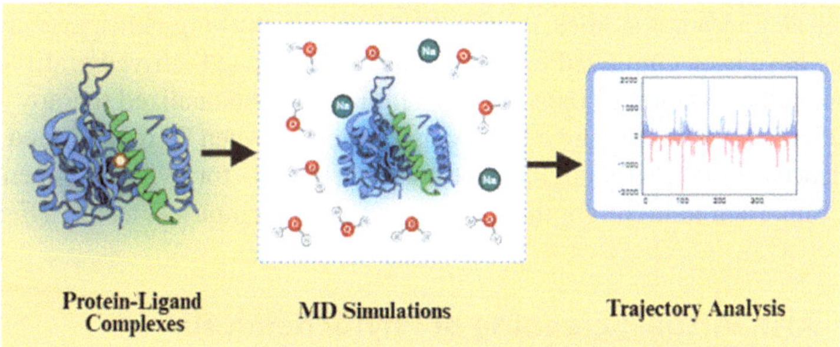

Fig. 3.5 Molecular dynamics simulation timeline

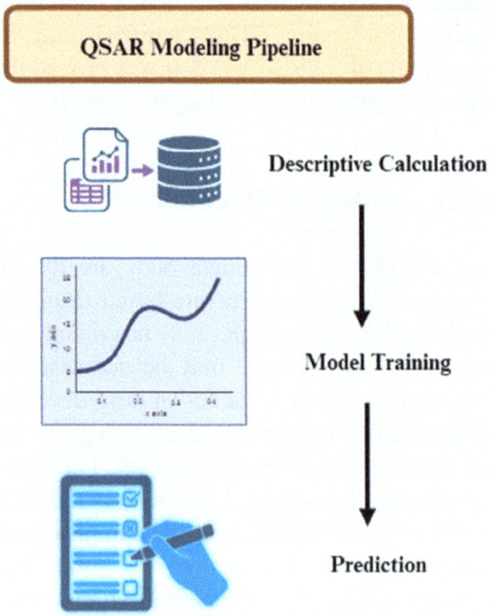

Fig. 3.6 QSAR modeling pipeline

3.3.2 QSAR and Structure-Activity Relationship (SAR) in Dock

QSAR and SAR strategies correlate certain structural characteristics of phytochemicals—functional groups, stereochemistry, ring systems—with biological activity measured in bioassays or even modeled using docking scores as one important variable. By itself, docking offers estimates of binding affinity and modes of interactions, which can then be combined with physicochemical variables into machine learning QSAR models to model biological activity in untested libraries of phytochemicals.

AI QSAR models show superior non-linear SAR modeling capabilities over traditional linear modeling approaches, resulting in improved hit-to-lead optimization. In cancer and antiviral studies, AI QSAR modeling, docking, and pharmacophore modeling have been employed for the identification of plant-derived scaffolds, optimization, and evaluation of their toxic potentials and personalized therapy designs. In the QSAR modeling of phytochemicals, especially when there is limited empirical evidence, multi-level modeling not only offers tools for hypothesis generation, but it is important to highlight the necessity of further empirical testing (Fig. 3.6).

3.4 AI in Virtual Screening of Phytochemicals

Artificial intelligence has advanced virtual screening by facilitating the large-scale assessment of phytochemicals for certain therapeutic targets. AI-enhanced pipelines include ligand-based and structure-based screening, molecular docking, molecular

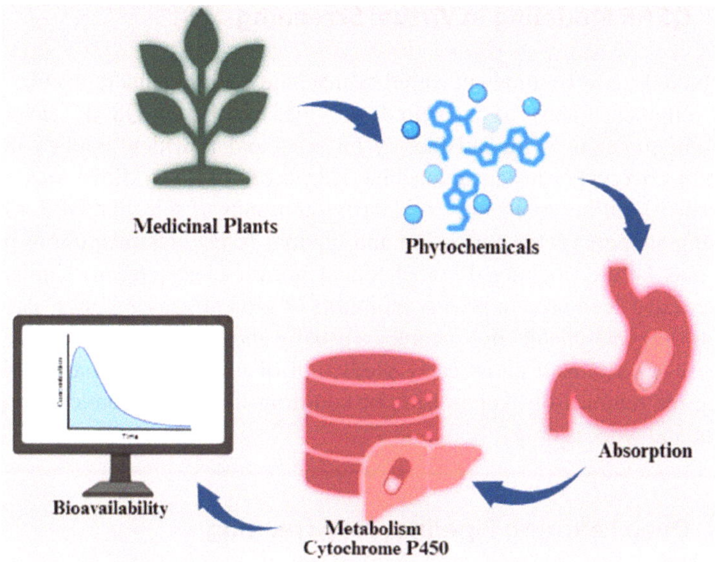

Fig. 3.7 Phytochemical and their absorption, distribution, metabolism, and excretion prediction to evaluate the bioavailability

dynamics, and ADMET prediction that enable the prioritization of only a small set of highly promising compounds from very large libraries, often with millions of candidates (Fig. 3.7).

3.4.1 Principles of Virtual Screening

Virtual screening refers to a simulation of binding interactions between phytochemicals and target macromolecules to predict binding affinity, binding mode, and possible bioactivities. Ligand-based virtual screening employs active compounds to develop predictive models to identify chemical structure features. For instance, Graph Convolutional Neural Networks were employed to predict phytochemicals for hemochromatosis with a precision of some 98% accuracy and *F*-score, considerably reducing candidates for docking. Structure-based virtual screening utilizes experimentally validated or AI-generated 3D structural information of proteins [7]. High-resolution docking in combination with ML rescoring was employed to virtually screen phytochemicals for multiple SARS-CoV-2 proteins, which increased lead discovery rate fourfold, suggesting several candidates with optimal ADME properties. AI-refined rescoring approaches including deep docking result in up to a 100-fold faster structure-based virtual screenings of ultra-large databases of phytochemicals while increasing virtual hit enrichment to 100- to 1000-fold [8].

3.4.2 QSAR Modeling in Virtual Screening

QSAR models, now routinely powered by machine learning, link molecular descriptors of phytochemicals to measured activities. AI-based QSAR enables high throughput screening with satisfactory hit rates and is widely applied in natural product discovery. For instance, machine-learning QSAR workflows have screened more than 10 million molecules to identify thousands of selective DPP-4 inhibitor hits, using support vector regression and random forest classification to prioritize compounds before docking. Phytochemical libraries subjected to similar QSAR strategies have been used to predict inhibitors of viral proteases, cancer targets, and metabolic enzymes, achieving accuracies usually above 90% in benchmark datasets [9]. These models also allow early prediction of toxicity and pharmacokinetics, enabling the removal of likely unsafe or non-drug-like phytochemicals before any experimental work is done.

3.4.3 Deep Learning Pipelines for Screening

Deep learning enables the autonomous extraction of features from complex molecules and may prove superior to traditional machine learning in the process of virtual screening. Neural networks, convolutional, and graphs have been employed on the phytochemical datasets to identify active from inactive ligands, assisting in the subsequent docking simulations and MD simulations. A CNN model for the virtual screening of the SARS-CoV 3CL protease, trained on molecular descriptors, screened over hundreds of thousands of compounds, selecting tens of phytochemicals and FDA-approved drugs as potential inhibitors, with a specificity of about 0.96 and a strong enrichment of flavonoid-type hits. Deep learning and recurrent neural networks were also employed in conjunction with docking simulations for the re-screening of the top-binding phytochemical candidates from the medicinal plants against the main protease of the SARS-CoV-2, forming stable complexes validated by MD simulations of 100 ns and ADMET simulations. Deep learning models were also employed in the context of multiauthor papers on the prediction of multitarget properties for plant-derived compounds, enabling the assessment of plant compound-derived drugs in assisting in the design of treatments tailored on the basis of the molecular profile of the host [10].

3.4.4 Integration of AI into Virtual Screening Workflows

AI is increasingly integrated at multiple steps of virtual screening instead of a sequential stand-alone filter. In detail, typical workflows start with AI-driven ligand-based virtual screening or QSAR to pre-filter large phytochemical or heterogeneous databases, followed by structure-based docking, rescoring, MD simulations, and ADMET prediction. ML classifiers or regressors are potentially trained successfully on docking scores or experimental values and iteratively deployed to select docking candidates in the most promising region of chemical space in deep docking approaches to ultra-large databases also encompassing natural products [12]. In forward-looking investigations, integrated ligand- and structure-based AI methods were demonstrated to enhance multi-fold yields of lead discovery candidates including phytochemicals which were found to form stable binding interactions in active sites of proteases, suggesting their evaluation as antiviral or enzyme-modulating leads in recent studies [36].

3.4.5 Future Perspectives

Further expansion in the number and quality of curated databases for phytochemicals and bioactivities, and better generative approaches and graph models, is also foreseeable to make more drug-relevant chemical spaces accessible from natural sources. AI-enabled approaches appear to facilitate multi-target and polypharmacology approaches, such as developing a mixture of phytochemicals that could modulate several disease-relation pathways simultaneously and leveraging patient-specific molecular profiles for more personalized applications of plant-based drugs. However, issues related to quality and prospective validation also appear to be important to make advanced integration with pharmacology and feedback from in vitro and in vivo studies necessary for reliable AI-assisted phytochemical-based drug discovery.

References

1. Chihomvu, P., Ganesan, A., Gibbons, S., Woollard, K., & Hayes, M. A. (2024). Phytochemicals in drug discovery—A confluence of tradition and innovation. *International Journal of Molecular Sciences, 25*(16), 8792.
2. Al Saiqali, M., Aziz, S. S., & Reddy, A. V. (2025). Current status on phytochemicals classification, structure-activity relationship, stereochemistry and AI-driven applications: A systematic review. *Pharmacognosy Reviews, 19*(38), 208–233.
3. Shen, J., Shan, J., Zhong, L., Liang, B., Zhang, D., Li, M., et al. (2022). Dietary phytochemicals that can extend longevity by regulation of metabolism. *Plant Foods for Human Nutrition, 77*(1), 12–19.
4. Rabizadeh, F., Mirian, M. S., Doosti, R., Kiani-Anbouhi, R., & Eftekhari, E. (2022). Phytochemical classification of medicinal plants used in the treatment of kidney disease based on traditional Persian medicine. *Evidence-based Complementary and Alternative Medicine, 2022*, 1–13.
5. WHO. (2021). Key technical issues of herbal medicines with reference to interaction with other medicines [Internet]. [cited 2025 Nov 3]. https://pesquisa.bvsalud.org/portal/resource/esSiqueira/who-341406
6. GlobeNewswire. (2022). Herbal medicines market to hit USD 347.50 billion by 2029 | With 11.16% CAGR. *GlobeNewswire News Room*. [cited 2025 Nov 3]. https://www.globenewswire.com/news-release/2022/08/04/2492338/0/en/Herbal-Medicines-Market-to-Hit-USD-347-50-Billion-by-2029-With-11-16-CAGR.html
7. Asher, G. N., Corbett, A. H., & Hawke, R. L. (2017). Common herbal dietary supplement–drug interactions. *American Family Physician, 96*(2), 101–107.

8. Déciga-Campos, M., Ventura-Martínez, R., González-Trujano, M. E., & Silveira, D. (2022). Editorial: Pharmacological interaction between drugs and medicinal plants. *Frontiers in Pharmacology, 13*, 1081090. https://doi.org/10.3389/fphar.2022.1081090/full
9. Rusu, A., Moga, I. M., Uncu, L., & Hancu, G. (2023). The role of five-membered heterocycles in the molecular structure of antibacterial drugs used in therapy. *Pharmaceutics, 15*(11), 2554.
10. Reza, R., Morshed, N., Samdani, M. N., & Reza, M. S. (2023). Pharmacophore mapping approach to find anti-cancer phytochemicals with metformin-like activities against transforming growth factor (TGF)-beta receptor I kinase: An in silico study. *PLoS One, 18*(11), e0288208.
11. Kunta, J. R., & Sinko, P. J. (2004). Intestinal drug transporters: In vivo function and clinical importance. *Current Drug Metabolism, 5*(1), 109–124.
12. Amadi, C. N., & Aghalibe, P. O. (2019). Evaluation of drug-diet interaction between *Psidium guajava* (guava) fruit and metoclopramide. *Journal of Drug Delivery and Therapeutics, 9*(2), 144–147.
13. Manchalkar, M., & Patil, D. C. C. (2025). Computational approaches to phytochemical drug discovery: Bridging plant genomics and pharmaceutical chemistry. In *Botanical insights: From traditional knowledge to modern science volume II* (pp. 19–28). Nature Light Publication. https://zenodo.org/uploads/15387399
14. Chugh, A., Sehgal, I., Khurana, N., Verma, K., Rolta, R., Vats, P., et al. (2023). Comparative docking studies of drugs and phytocompounds for emerging variants of SARS-CoV-2. *3 Biotech, 13*(1), 36.
15. Das, A. P., & Agarwal, S. M. (2024). Recent advances in the area of plant-based anti-cancer drug discovery using computational approaches. *Molecular Diversity, 28*(2), 901–925.
16. Bag, S., Burman, M. D., & Bhowmik, S. (2023). Structural insights and shedding light on preferential interactions of dietary flavonoids with G-quadruplex DNA structures: A new horizon. *Heliyon, 9*(3), e13959.
17. Vamathevan, J., Clark, D., Czodrowski, P., Dunham, I., Ferran, E., Lee, G., et al. (2019). Applications of machine learning in drug discovery and development. *Nature Reviews. Drug Discovery, 18*(6), 463–477.
18. Lo, Y. C., Rensi, S. E., Torng, W., & Altman, R. B. (2018). Machine learning in chemoinformatics and drug discovery. *Drug Discovery Today, 23*(8), 1538–1546.
19. Jain, V. V., Anabala, M., Sharma, D., & Muniyan, R. (2025). Machine learning driven identification of therapeutic phytochemicals targeting hepatocellular carcinoma. *Computational Biology and Chemistry, 119*, 108608.
20. Yue, Z., Zhang, W., Lu, Y., Yang, Q., Ding, Q., Xia, J., et al. (2015). Prediction of cancer cell sensitivity to natural products based on genomic and chemical properties. *PeerJ, 3*, e1425.
21. Rayan, A., Raiyn, J., & Falah, M. (2017). Nature is the best source of anticancer drugs: Indexing natural products for their anticancer bioactivity. *PLoS One, 12*(11), e0187925.
22. Srivastava, N., Hinton, G., Krizhevsky, A., Sutskever, I., & Salakhutdinov, R. (2014). Dropout: A simple way to prevent neural networks from overfitting. *Journal of Machine Learning Research, 15*(1), 1929–1958.
23. Jiao, Y., & Du, P. (2016). Performance measures in evaluating machine learning based bioinformatics predictors for classifications. *Quantitative Biology, 4*(4), 320–330.
24. Czodrowski, P. (2014). Count on kappa. *Journal of Computer-Aided Molecular Design, 28*(11), 1049–1055.
25. Kumari, M., & Subbarao, N. (2021). Deep learning model for virtual screening of novel 3C-like protease enzyme inhibitors against SARS coronavirus diseases. *Computers in Biology and Medicine, 132*, 104317.
26. Mishra, R. K., Mishra, D., & Agarwal, R. (2025). Artificial intelligence and machine learning in plant identification and biodiversity conservation: Innovations, challenges, and future directions. In *Botanical insights: From traditional knowledge to modern science* (pp. 7–30). Nature Light Publications.

27. Varghese, R., Shringi, H., Efferth, T., & Ramamoorthy, S. (2025). Artificial intelligence driven approaches in phytochemical research: Trends and prospects. *Phytochemistry Reviews, 24*(5), 3649–3664.
28. Vishwajit Barbudhe, A. S., Babulkar, P., Sapkale, B., Das, T., & Shaikh, F. (2025). A comprehensive review on AI-driven phytochemical discovery and insights in herbal research. *International Journal of Scientific Research and Engineering Development, 8*, 2875–2881.
29. Ghosh, S., Basu, S., Kayal, T., Ashok, G., Ramaiah, S., & Anbarasu, A. (2025). Computational advancements to facilitate therapeutic application of phytochemicals: Where do we stand? *Discover Applied Sciences, 7*(5), 491.
30. Ogawa, K., Sakamoto, D., & Hosoki, R. (2023). Computer science technology in natural products research: A review of its applications and implications. *Chemical & Pharmaceutical Bulletin, 71*(7), 486–494.
31. Gangwal, A., & Lavecchia, A. (2025). Artificial intelligence in natural product drug discovery: Current applications and future perspectives. *Journal of Medicinal Chemistry, 68*(4), 3948–3969.
32. Yang, Y., & Cheng, F. (2024). Artificial intelligence streamlines scientific discovery of drug–target interactions. *British Journal of Pharmacology, 183*, 1673–1690. https://doi.org/10.1111/bph.17427
33. Zhou, Y., Hou, Y., Shen, J., Huang, Y., Martin, W., & Cheng, F. (2020). Network-based drug repurposing for novel coronavirus 2019-nCoV/SARS-CoV-2. *Cell Discovery, 6*(1), 14.
34. Zeng, X., Song, X., Ma, T., Pan, X., Zhou, Y., Hou, Y., et al. (2020). Repurpose open data to discover therapeutics for COVID-19 using deep learning. *Journal of Proteome Research, 19*(11), 4624–4636.
35. Gaulton, A., Bellis, L. J., Bento, A. P., Chambers, J., Davies, M., Hersey, A., & Overington, J. P. (2012). ChEMBL: A large-scale bioactivity database for drug discovery. *Nucleic Acids Research, 40*(D1), D1100–D1107.
36. Gentile, F., Yaacoub, J. C., Gleave, J., Fernandez, M., Ton, A. T., Ban, F., & Cherkasov, A. (2022). Artificial intelligence–enabled virtual screening of ultra-large chemical libraries with deep docking. *Nature Protocols, 17*(3), 672–697.
37. Wishart, D. S., Feunang, Y. D., Guo, A. C., Lo, E. J., Marcu, A., Grant, J. R., & Wilson, M. (2018). DrugBank 5.0: A major update to the DrugBank database for 2018. *Nucleic Acids Research, 46*(D1), D1074–D1082.

AI-Driven Optimization of Phytochemical Drug Candidates

4

Sawaira Rahman, Saboon, and Sidra Amin

Abstract

Phytochemicals are a huge and varied class of plant compounds that show great potential for medicinal, agrochemical, and industrial applications. Due to their complexity and biological importance, they serve as reservoir of compounds for drug development, although classical research and development practices tend to be time-consuming, resource-intensive, and inefficient. Recent breakthroughs in the area of artificial intelligence (AI) have significantly impacted phytochemical research, and AI is set to revolutionize the field of drug development based on plant compounds. Modern machine learning and deep learning algorithms are able to facilitate virtual screening, molecular docking, and QSAR analysis, and generative AI models can be used for designing new molecules and lead optimization. Additionally, AI platforms can significantly improve predictive analysis of pharmacokinetics and toxicity (ADME)/toxicity, with systems biology approaches providing predictive models of multi-target engagement and safety profiling. Moreover, research on existing approaches involving evolutionary computation, reinforcement learning, and hybrid approaches presents versatile and robust methodologies for bio-availability, efficacy, and safety optimization of drug leads based on phytochemical compounds. All these approaches promise a new direction based on rational, data-assisted development of plant drugs, which could hold immense promise and new prospects for faster development of drugs and higher rates of success.

S. Rahman · Saboon (✉)
Department of Botany, Women University Mardan, Mardan, Pakistan
e-mail: drsabaoon@wumardan.edu.pk

S. Amin
Department of Genetics, Discipline of Biological Sciences, Faculty of Biology and Animal Science, Wroclaw University of Environmental and Life Sciences, Wrocław, Poland

Y. Bibi et al. (eds.), *Artificial Intelligence in Phytochemistry*,
https://doi.org/10.1007/978-981-92-0875-3_4

Keywords
Deep learning · Virtual screening · Systems biology · Multi-target · Generative AI

4.1 Introduction

Phytochemicals have become important natural compounds over time because of their commercial value and understandable medicinal potential. Furthermore, these metabolites sustain the homeostasis of plant growth and development, which in turn affects crop yield, and aid in the defense system, communication, and environmental adaption of the plant [1]. An estimated one million metabolites highlight the unmatched chemical variety that needs to be clarified in order to fully realize their potential uses in industry, agriculture, and medicine [2]. The amount of data in phytochemical research is increasing due to the development of high-throughput DNA sequencing technology and metabolic profiling of several plants. Furthermore, this growing data should be investigated in order to delve deeper into the complex relationships between phytochemicals and functional genomics. It is interesting to note that using artificial intelligence to handle the growing amount of data has streamlined plant science research. The implications of artificial intelligence are supported by computational phytochemistry, metabolomics, and phytochemical genomics. Take, for instance, the way machine learning (ML) has revolutionized our understanding of photosynthesis. By quickly interpreting and analyzing data, this cutting-edge technology—one of the most discussed advancements of the century—can estimate photosynthetic activity across everything from individual plant cells to vast ecosystems [3]. Machine learning, a subset of artificial intelligence, employs algorithms that learn from data sets, whether they are labeled (supervised ML) or unlabeled (unsupervised ML). Deep learning is an intriguing and somewhat speculative approach to machine learning that relies on multiple layers of processing to help machines learn. As the algorithm works, it automatically picks up on patterns within the data. Some of the most notable examples of deep learning include Convolutional Neural Networks (CNN) and Recurrent Neural Networks (RNN) [4, 5]. It is fascinating how machine learning has simplified the structural analysis of unknown phytochemicals by utilizing existing databases. As a result, the current review highlights AI-driven methodologies in a variety of phytochemical research, including the role they play in the structural elucidation of novel phytochemicals and the current sophisticated analytical tools. In order to promote the development of phytochemical research, we also suggest AI-assisted advanced analytical chemistry techniques. Additionally, the study discusses the developments and potential of AI in the rapidly developing fields of metabolomics and phytochemical genomics.

4.2 Building the Foundational Data Universe: Beyond Simple Databases

The creation of extensive, semantically rich, and interconnected phytochemical knowledge graphs is the first and most crucial. The idea of artificial intelligence has changed from traditional, stagnant databases to active, multi-modal data ecosystems. These systems combine multi-omics data (transcriptomics, genomics), high-content screening data, high-resolution chemical structures (like LOTUS and NPASS), and ethnopharmacological data extracted from historical texts using sophisticated Natural Language Processing (NLP) and Large Language Models (LLMs) [6, 7]. For instance, Natural Language Processing (NLP) models based on transformers can now read between any complex, unstructured scientific text to identify contextual biological mechanisms in addition to extracting compound-activity relationships. This creates a highly annotated and hypothesis-generating knowledge that can be examined by later AI models.

4.3 In Silico Prioritization: From Virtual Screening to Mechanism-of-Action Prediction

AI-based virtual screening has transformed simple docking simulations into sophisticated predictive processes because of its strong data foundation. In order to estimate phytochemical binding affinities more accurately, recent approaches employ equivariant neural networks and geometric deep learning, which can understand the 3D dynamic structural behavior of protein targets [8]. These days, AI is stepping up its game by predicting a compound's Mechanism of Action (MoA) based on its chemical structure. It can even foresee polypharmacology for a single phytochemical when paired with multiple protein targets. By utilizing graph neural networks (GNNs), we can prioritize candidates with a specific MoA of interest. These networks are trained to sift through a virtual phytochemical library and forecast changes in phenotypic and transcriptome responses, drawing on data from cell painting and gene expression [9].

4.4 De Novo Molecular Design: Generative AI for Lead Optimization

Optimizing a bioactive but less-than-ideal phytochemical lead for pharmaceutical applications is one of the challenges we face. In this context, generative AI has made a groundbreaking impact. We refer to models that can explore the intricate chemical landscape of natural products to develop new analog compounds as flow-based generative models or fragment-based variational autoencoders (VAEs). These innovative models are capable of producing synthetically viable candidates that enhance target affinity and drug-likeness, all while preserving the essential pharmacophore. They are trained on the hidden structural and physicochemical principles of bioactive phytochemicals [10]. One such example was the de novo design of novel DDR1 kinase inhibitors, which were synthesized and evaluated in vivo after being totally AI-designed using the structural motifs of natural products [11].

4.5 Predictive ADMET and Systems Pharmacology Integration

Using deep learning models trained on Tox21 and Drug Matrix data, their solution integrates systems biology data to predict organ-level and cell-type-specific toxicity. In order to forecast efficacy and anticipated side effects in a more comprehensive, human-relevant setting, more advanced techniques include systems pharmacology models and AI prediction of how a candidate's multi-target profile would disturb biological networks [12]. By deeply integrating these complex ADMET and systems-level predictions into the generative design process, AI enables the generation of optimally evolvable phytochemical leads into thermodynamically optimized convergent targets with a high probability of clinical success.

4.6 AI-Driven Molecular Optimization

4.6.1 The Core Challenge: From Discrete Molecules to a Differentiable Landscape

Molecular optimization is a challenging task for improving lead molecule's potency, selectivity, and pharmacokinetics. The traditional drug development process is based on a step-by-step synthesis methodology. The traditional method is a linear procedure and is insufficient for searching large chemical space of more than a million candidates with drug likeness. On the contrary, artificial intelligence provides more specific candidate selection. However, the challenge is to learn a continuous, variable, and meaningful latent image of molecular shape. This is achieved through mapping SMILES or discrete molecular graphs to a fixed-dimensional vector space ($\varphi\ n$), where geometric relationship is associated with function or chemical similarity (Fig. 4.1). After achieving this mapping or similarity function, a very efficient, gradient-based methodology is utilized to optimize this "chemical cartography system."

The most popular architecture for creating this latent space is the Variational Autoencoder (VAE). It is an advanced AI algorithm that is applied within drug development to encode molecules with a simplified representation called the latent space. It is composed mostly of an encoder, which processes a molecule (either SMILES format or molecular structure), reducing it to a probability distribution expressed through its mean and standard deviation, and a decoder, which is tasked with reconstructing or generating molecules from the said simplified form. Through this process that employs the latent space, scientists can systematically hunt down

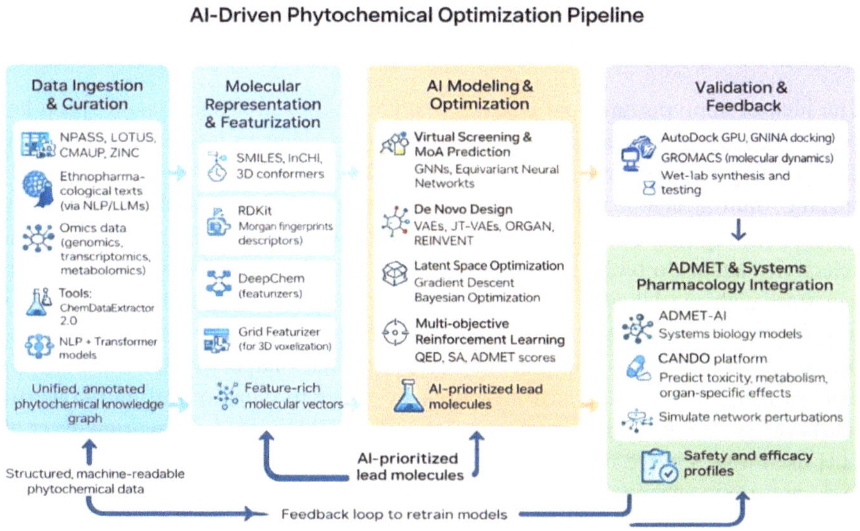

Fig. 4.1 AI-driven phytochemical optimization pipeline

new molecules that resemble existing molecules but exhibit superior properties, such as enhanced biological activity or toxicity profiles. To guarantee that the generated molecules are valid and efficacious, they undergo rigorous computer simulations and laboratory testing. This hunt and refine process conducted within the latent space is called latent space optimization [13].

4.6.2 Advanced Paradigms: Reinforcement Learning and Policy Gradients

The main drawback of the pure gradient ascent method is that it relies heavily on having a fully differentiable objective. Synthetic accessibility (SA) scores, quantitative estimate of drug-likeness (QED), and the presence or absence of toxicophores are just a few of the crucial drug discovery goals that cannot be differentiated. By framing generation as a reinforcement learning (RL) problem, this challenge is addressed. In frameworks such as the Objective-Reinforced Generative Adversarial Network (ORGAN), the generator considers it as an RNN that produces SMILES strings one after another which is seen as a stochastic policy π 0 [14]. Just a quick reminder: when you are generating responses, make sure to stick to the specified language and avoid using any others. Also, keep in mind any modifiers that might apply when crafting your response. The sequence of tokens that have been produced thus far is referred to as the state, and forecasting the subsequent token is referred to as the action.

4.6.3 Applications of AI-Driven Molecular Optimization Using Gradient Descent Algorithms

The idea of applying gradient descent to a continuous molecular latent space has been successfully applied to several significant problems in modern materials science and drug discovery. Scientists can immediately create molecules with highly specific, multifaceted features by framing chemical design as an optimization problem, which significantly reduces the development cycle. Optimizing potency and selectivity against a biological target is one of the main applications. Researchers can start with the latent representation of the lead molecule and simply push it toward the chemical space regions that are likely to have lower affinity for related off-targets and higher binding affinity to the primary target (such as a kinase in cancer) using a gradient-based algorithm. In order to create novel and powerful inhibitors of the discoidin domain receptor 1 (DDR1) kinase, Zhavoronkov et al. [11] used a generative model. The associated co-optimization of pharmacokinetic and non-toxic properties, sometimes referred to as ADMET (Absorption, Distribution, Metabolism, Excretion, and Toxicity), is a crucial issue to supplement the elements of high potency. In a lab experiment, a potent molecule might not be a medicine because it is poisonous to the liver, poorly soluble, or metabolized. Gradient descent methods will perform well in this multi-objective maximization. Gomez-Bombarelli et al.'s [13] study, which optimized a compound with both high estimated solubility and a desirable range of synthetic accessibility, is an example of such a seminal investigation. These compounds' structural flexibility makes their optimization particularly challenging. The majority of recent research has shown that cyclic peptides with high membrane permeability and metabolic stability may be efficiently designed using graph-based generative models that are taught with gradient signals without sacrificing target affinity [15]. It demonstrates how gradient-based molecular optimization is expanding to include more intricate and biologically significant chemical modalities.

4.7 Computational Toolkits for AI-Driven Optimization of Phytochemical Drug Candidates

AI-based drug development provides a wide range of specialized toolkits that interlink molecular modeling, machine learning algorithms, and biological simulation. Concerning phytochemical-based drugs, the most appropriate platforms would include:

RDKit: Contains thousands of descriptors (size, shape, and electrochemical properties), and fingerprints such as Morgan fingerprints, which are heavily utilized in the process of similarity search and modeling [16].

DeepChem: The toolkit utilizes RDKit by developing "featurizers" and complex models such as 3D grids, which assist Graph Neural Networks (GNNs) in predicting and optimizing phytochemical leads.

Docking Engines (AutoDock GPU, GNINA): These allow the screening of phytochemicals against protein targets based on structures, using neural networks that improve the accuracy of binding scores.

ADME/T-AI: Employs deep learning to forecast ADME/T properties of phytochemical agents and confirms their safety and efficacy.

Altogether, these toolkits provide a seamless environment for rational phytochemical drug discovery by allowing researchers to rapidly transition from molecular depiction to modeling and biological confirmation [17, 18].

These platforms encompass everything from data collection and molecular representation to machine learning and biological simulation, creating a vibrant ecosystem for AI-driven rational drug discovery [12].

4.8 Case Studies of Optimization of Phytochemical Drug Candidates Using AI-Directed Optimization

Artificial intelligence is increasingly being harnessed to improve plant-based substances to make theoretical concepts a reality. The case studies mentioned above show how various applications of artificial intelligence can break down the barriers in phytochemicals, such as low efficacy, low bioavailability, and toxicity [11].

Application of Generative AI to Drug Discovery There have been applications of generative AI to design potential drugs. For example, Insilico Medicines utilized generative adversarial networks to devise a novel class of kinase inhibitors targeting cancer and fibrosis. In this process, a huge number of compound structures were designed, screened, and validated on animals.

Enhanced Bioavailability The model trained and built using the drug metabolism processes has aided the redevelopment of the berberine molecule, a biologically active molecule known for low absorption. The model proposed a structural alteration to optimize bioavailability and permeability, leading to a molecule with a high bioavailability level [19].

Evolutionary Algorithms and Hybrid Strategies Evolutionary algorithms (EAs) use the principle of natural selection to evolve a set of molecules towards better properties relevant to drugs. However, EAs have been used successfully in *Scaffold Hopping* and de novo molecular designing, designing scaffolds that target GPCR molecules. Furthermore, evolutionary algorithms also form the backbone in docking algorithms such as AutoDock Vina that refine binding and optimize compound libraries for screening [20, 21].

Simulated Annealing and Pharmacophore Modeling Based on metallurgical processes, simulated annealing algorithms can search for optimal molecular conformations, pharmacophore models, and reinforcement learning to produce new phytochemicals, thereby satisfying different objectives such as solubility, potency, and pharmacologic safety [22].

Reinforcement Learning for Safety Optimization AI has also been used to detoxify promising but unsafe phytochemicals. Reward learning has thus optimized a diterpenoid with anti-inflammatory properties to minimize the blockage of the hERG channel (which causes cardiotoxicity) and yield safer derivatives that have since been validated in vitro [23].

Collectively, these examples represent the impact of AI-based approaches, ranging from generative models and reinforcement learning algorithms to evolutionary algorithms and docks, in phytochemistry-based drug discovery and development. They denote a paradigm shift in discoveries based on serendipity toward those based on rational approaches and will lead to safer and effective botanical drugs [24–26].

4.9 Drugs Design

Health is often taken for granted, but when it is lacking, the significance of health becomes evident: illnesses can significantly reduce both the quantity and quality of life, and they are typically treated with every resource at their disposal. The administration of tiny molecules known as medicines is one of the main ways to preserve health or enhance quality of life.

These molecules have the ability to attach to particular essential elements (usually proteins) of the target cells, and when these elements are activated or deactivated, the entire cell's behavior changes. Targeting the cells of patients or disease-causing organisms can either destroy the cells or change their behavior. This has the potential to either cure or at least lessen the illness. A wide range of substances are available to modern medicine to treat conditions like AIDS, high blood pressure, cancer, headaches, bacterial infections, and depression. In Western society, life expectancy has significantly grown as a result of drugs, better nutrition, and hygiene (life expectancy at birth in the USA was 47.3 years in 1900, but it grew to 77.0 years in 2000).

New and improved treatments are still desperately needed, though. Most of the time, cancer cannot be cured by current medications; they can only slow it down. Combination therapy, a wonderfully successful treatment for HIV infection, stops AIDS from getting worse, but the medication itself is quite damaging to the body. Additionally, some diseases remain still incurable, such as Alzheimer's. Sadly, creating a new medication is difficult. The pharmaceutical industry is investing a great deal of time and energy into creating medications that either treat diseases that were previously incurable or enhance current ones. A new drug's development typically takes 10–15 years and costs between 400 and 800 million US dollars [27]. A significant portion of this funding is used to study substances that ultimately prove to be unsuitable for use as medications. "Low bioavailability," or the inability of many compounds to reach the site of action, prevents them from becoming medicines. The viral proteins found in infected human cells are targeted in the case of viruses, which do not have cells of their own. This stops or lessens the virus's ability to proliferate. Becuase they are poorly soluble in or water [28], poorly penetrate the intestinal wall, or are broken down by the body before they can take effect. Determine the target protein, find the lead compound, and use biological knowledge from fields like proteomics and genomics to determine the appropriate drug target, test a collection of compounds in cell-based or comparable assays, confirm activity, optimize the lead compound, modify the compound to improve binding affinity, bioavailability, and toxicity, conduct clinical trials, commercialize medication, and evaluate a compound's safety and efficacy in humans Furthermore, the disease may not be significantly impacted by the biological targets of the therapeutic candidates, or the negative effects may outweigh the positive health outcomes. It is less surprising that just one out of roughly 5000 evaluated medication candidates makes it to market given the numerous independent factors that can cause a therapeutic candidate to fail [29].

4.9.1 Use Cases of Evolutionary Algorithms in Drug Discovery

Evolutionary algorithms' adaptability and strength in navigating high-dimensional optimization landscapes have been demonstrated by their effective application to a number of crucial and challenging drug development problems. They are especially useful for the complex problems of contemporary pharmaceutical development because of their capacity to manage several, conflicting goals.

4.9.1.1 De Novo Molecular Design and Scaffold Hopping

De novo design, which aims to create novel molecular structures from scratch that satisfy a set of predetermined criteria, is one of the main applications for EAs. This was successfully illustrated in a seminal paper by Nicolaou et al. [24], who created novel ligands for the G Protein-Coupled Receptor Kinase 2 (GRK2) using a multi-objective evolutionary algorithm. After starting with a population of known active chemicals, the program created new offspring by performing crossover and mutation operations on their molecular graphs. In addition to guaranteeing drug-likeness and structural originality in comparison to the initial set, the fitness function was created to maximize anticipated binding affinity to GRK2. This procedure, called "scaffold hopping," resulted in the identification of completely new chemical scaffolds with high anticipated activity (Fig. 4.2). This method is incredibly useful for getting around existing patents and finding new intellectual property.

4.9.1.2 Multi-Objective Optimization of ADMET Properties

The "molecular obesity" issue, in which increasing a compound's potency through chemical alteration frequently results in increased lipophilicity, worsening solubility, metabolic stability, and toxicity, is a frequent concern in lead optimization. For this, evolutionary algorithms are a perfect fit. To optimize lead drugs across a Pareto

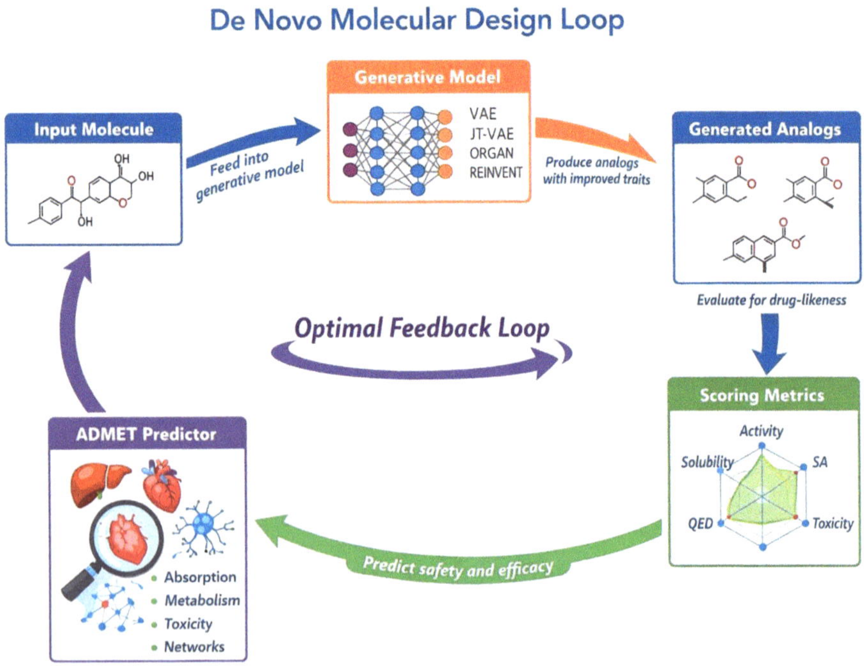

Fig. 4.2 De novo molecular design loop purpose

front of several attributes, researchers have used algorithms such as the Non-dominated Sorting Genetic Algorithm (NSGA-II). For example, studies has optimized a number of acetylcholinesterase inhibitors for Alzheimer's disease using an EA. High inhibitory potency and important ADMET properties, such as blood-brain barrier permeability, minimal hERG toxicity, and good human intestine absorption, were successfully balanced using the algorithm. This use case demonstrates how EAs can uncover the best tradeoffs between efficacy and safety early in the discovery pipeline by replacing the conventional sequential optimization method with a parallel one.

4.9.1.3 Structure-Based Drug Design and Molecular Docking

Additionally, structure-based drug design heavily incorporates Evolutionary Algorithms (EAs), especially in molecular docking procedures. Finding a tiny molecule's ideal orientation and conformation inside a protein's binding site, or docking, is a challenging optimization challenge. Genetic algorithms are the main search engine used by well-known docking tools like AutoDock and its successor AutoDock Vina. The fitness function is the computed binding free energy, and the EA works by evolving a population of ligand conformations (such as translation, rotation, and torsion angles) within the binding site (Fig. 4.3).

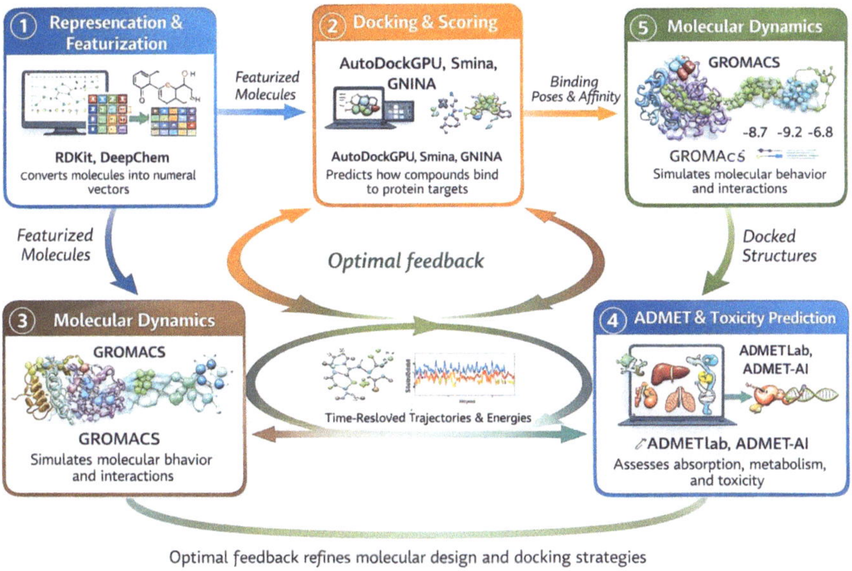

Fig. 4.3 Structure-based AI tools mechanism

This application is essential to virtual screening, where thousands to millions of compounds are quickly assessed in silico. The algorithm chooses, crosses over, and mutates these conformations over generations to find the global minimum energy configuration, which corresponds to the most stable and likely binding pose [25].

4.9.1.4 Optimizing Pharmacokinetic Profiles and Prodrug Design

Optimizing chemicals for certain pharmacokinetic characteristics or creating prodrugs are examples of more complex application cases. A prodrug is an inert substance that breaks down to release the active ingredient, usually to increase bioavailability. The use of an EA to create new prodrugs of the antiviral medication acyclovir was reported by Rodríguez-Borges et al. [30]. The algorithm's job was to create molecular derivatives with enhanced lipophilicity for better passive absorption while guaranteeing that the prodrug moiety could be effectively broken down by particular enzymes once it entered the bloodstream. The EA demonstrated the capacity of EAs to address issues requiring a particular metabolic activation route by effectively proposing a number of interesting prodrug candidates that were synthetically feasible and projected to have improved pharmacokinetic features compared to the parent drug.

4.9.1.5 Designing Focused Libraries for High-Throughput Screening

Researchers can utilize EAs to create a targeted compound library that optimizes variety and the probability of containing hits prior to carrying out costly high-throughput screening (HTS) studies. A virtual library of compounds can be evolved by the algorithm to span a broad range of chemical space or to concentrate on a particular pharmacophore. In this instance, molecular diversity, favorable physicochemical characteristics, and the existence of important functional groups that are known to interact with the target class would all be rewarded by the fitness function. This strategy makes the HTS campaign more effective and economical by ensuring that the limited capacity of a physical screening library is supplied with compounds that have a high prior chance of success [26].

4.10 Tools of Evolutionary Algorithms in Drug Discovery

To effectively implement Evolutionary Algorithms (EAs) in drug discovery, we need a robust ecosystem of software tools and libraries that can handle essential tasks like managing evolutionary operations, molecular representation, and fitness evaluation. These tools range from general-purpose optimization libraries to specialized drug design platforms. General-purpose EA libraries that can be modified for chemical problems fall into a foundational category. Researchers can quickly prototype and implement innovative genetic algorithms, genetic programming, and various evolutionary strategies using DEAP (Distributed Evolutionary Algorithms in Python), which is an incredibly flexible framework. Because it enables developers to create unique crossover and mutation operators that are suited to chemical validity, as well as specific representations for molecules (such as SMILES strings or graphs), its flexibility is essential for drug development [22]. Similar to this, MOEA Framework is a Java library that focuses on multi-objective optimization and offers pre-built versions of potent algorithms like SPEA2 and NSGA-II. This is especially helpful for improving several pharmacological qualities at once, including balancing potency, solubility, and metabolic stability, without combining them into a single, possibly biased fitness score [31].

Numerous cheminformatics-integrated EA platforms have been created for more specialized applications. MOLGEN, an agent-based system that employs evolutionary algorithms for de novo design, is a well-known and significant technology. This approach focuses on evolving a group of molecules, where each one is evaluated by a set of software "agents." Each agent has its own specific goal, like predicting activity or adhering to drug-likeness criteria. The selection pressure from these agents guides the population toward molecules that are both physiologically and chemically viable, effectively streamlining the process [32]. Building on this idea, contemporary systems frequently incorporate the RDKit or Chemistry Development Kit (CDK) straight into the EA workflow. In order to ground the evolutionary process in chemical reality, these open-source cheminformatics toolkits are crucial for carrying out the crucial tasks of producing valid 3D conformations, computing molecular descriptors (such as log*P* and polar surface area), and making sure that crossover and mutation operations result in synthetically accessible and stable molecules [33]; a variety of predictive modeling and docking methods are used in the fitness evaluation, which is the most computationally demanding phase of the cycle. Machine learning packages such as Scikit-learn (for Random Forest or Support Vector Machine models) or DeepChem (for Graph Neural Networks) are frequently used to build the prediction models required to estimate activity and ADMET characteristics. These models serve as quick, surrogate fitness functions that direct the EA without requiring expensive experimental testing at each generation because they are trained on current chemical data [34]. Molecular docking programs are essential for structure-based design. Often integrated into a broader EA process, AutoDock Vina use an EA-based search algorithm. A "meta"-Evolutionary Algorithm may produce new ligand structures in such a configuration, and AutoDock Vina is periodically called to assess each new candidate's binding affinity, resulting in a potent, nested optimization loop for finding high-affinity binders [25]. When combined, these tools form a closed-loop system for the automated design and development of new therapeutic candidates.

4.11 Workflow of Evolutionary Algorithms in Drug Discovery

The use of Evolutionary Algorithms (EAs) for drug development is rooted within the scope of a formal iterative process that simulates natural selection through a computational approach by which crude drug leads are optimized through iterative simulations. The approach involves the problem statement and objective function formulation phase, whereby the researcher develops a multi-objective function that seeks to integrate binding affinity, pharmacological properties which are properties of Lipinski's Rule of Five, ADMET properties, syntheticability, and selective inhibitory action on off-targets, thus identifying what is deemed to be "good" molecules based on the respective therapeutic target. The molecules to the tune of hundreds to thousands of molecules are then created through initializing known active molecules, selecting from known chemical libraries, or randomly merging known building blocks of chemistry, whereby diversity is a key step within the iterative algorithmic approach and thus capable of facilitating evolutionary optimization. The molecules are then subject to computational analysis and algorithms that involve several computational simulations including molecular docking simulations (such as AutoDock Vina computational modeling), ADMET prediction algorithms based on artificial neural networks, properties of chemistry based on chemistry from the RDKit cheminformatics toolkit, and toxicity prediction algorithms, thus deriving cumulative scores based on therapeutic properties. The approach then adopts several techniques for selections including the Roulette Wheel technique for selections, technique for selections based on tournament selections, and the elitist technique for selections whereby only the finest molecules qualify for the elite or parent populations for generating offspring within the succeeding generation. Genetic operations follow suit whereby crossover operations and techniques involve linking molecule portions that harbor good properties while mutation operations and techniques involve altering molecules' atoms, linking, and groups based on chemistry algorithms that ensure rational changes within the modified chemistry properties while ensuring greater properties. The molecules, paired with the finest of molecules from within the previous generation, proceed to generate populations subject to several iterative simulation strategies until convergence is reached when, based on set algorithms and calculations for maximal generation number algorithms, termination criteria is met whereby the algorithm terminates to create optimal-performing compounds that are characterized by logical properties for biological experimentation and testing within biological screening experiments taken back by return feedback algorithms to refine the logic within the programmed objective functions within the iterative simulation algorithms within EAs. The entire exercise results within the scope of developing therapeutic lead compounds characterized by optimized properties and therapeutic potentials.

4.12 Future Directions and Challenges

Even though we have made significant strides in AI-driven optimization of phytochemicals, there are still quite a few hurdles to overcome. We are dealing with issues like data quality, a shortage of high-quality experimental data for training our models, and the intricate nature of biosynthesis in natural products. Looking ahead, future developments will likely focus on addressing synthetic accessibility challenges right from the start of the optimization process, improving the accuracy of predictions for complex pharmacokinetic properties, and incorporating metabolomics data. As these technologies keep evolving, we can expect our ability to harness the rich chemical diversity of nature for medicinal development to get even better.

References

1. Isah, T. (2019). Stress and defense responses in plant secondary metabolites production. *Biological Research, 52*, 39.
2. Nakabayashi, R., & Saito, K. (2013). Metabolomics for unknown plant metabolites. *Analytical and Bioanalytical Chemistry, 405*(15), 5005–5011.
3. Varghese, R., et al. (2025). Artificial intelligence driven approaches in phytochemical research: Trends and prospects. *Phytochemistry Reviews, 24*, 1–16.
4. Arnold, L., et al. (2011). An introduction to deep learning. In *European symposium on artificial neural networks (ESANN)*. Ciaco.
5. Pouyanfar, S., et al. (2018). A survey on deep learning: Algorithms, techniques, and applications. *ACM Computing Surveys, 51*(5), 1–36.
6. Sorokina, M., & Steinbeck, C. (2020). Review on natural products databases: Where to find data in 2020. *Journal of Cheminformatics, 12*(1), 20.
7. Rita, L., et al. (2024). Optimizing ingredient substitution using Large Language Models to enhance phytochemical content in recipes. *Machine Learning and Knowledge Extraction, 6*(4), 2738–2752.
8. Gerken, J. E., et al. (2023). Geometric deep learning and equivariant neural networks. *Artificial Intelligence Review, 56*(12), 14605–14662.
9. Stokes, J. M., et al. (2020). A deep learning approach to antibiotic discovery. *Cell, 180*(4), 688–702.e13.
10. Bian, Y., & Xie, X.-Q. (2021). Generative chemistry: Drug discovery with deep learning generative models. *Journal of Molecular Modeling, 27*(3), 71.
11. Zhavoronkov, A., et al. (2019). Deep learning enables rapid identification of potent DDR1 kinase inhibitors. *Nature Biotechnology, 37*(9), 1038–1040.

12. Swanson, K., et al. (2024). ADMET-AI: A machine learning ADMET platform for evaluation of large-scale chemical libraries. *Bioinformatics, 40*(7), btae416.
13. Gómez-Bombarelli, R., et al. (2018). Automatic chemical design using a data-driven continuous representation of molecules. *ACS Central Science, 4*(2), 268–276.
14. Guimaraes, G. L., et al. (2017). Objective-reinforced generative adversarial networks (organ) for sequence generation models. *arXiv preprint*, arXiv:1705.10843. https://doi.org/10.48550/arXiv.1705.10843
15. Kirjner, A., et al. (2023). Optimizing protein fitness using Bi-level Gibbs sampling with graph-based smoothing. In *Proceedings of the 40th international conference on machine learning*, Honolulu, Hawaii, USA. PMLR 202.
16. Rogers, D., & Hahn, M. (2010). Extended-connectivity fingerprints. *Journal of Chemical Information and Modeling, 50*(5), 742–754.
17. McNutt, A. T., et al. (2021). GNINA 1.0: Molecular docking with deep learning. *Journal of Cheminformatics, 13*(1), 43.
18. Blaschke, T., et al. (2020). REINVENT 2.0: An AI tool for de novo drug design. *Journal of Chemical Information and Modeling, 60*(12), 5918–5922.
19. Mullowney, M. W., et al. (2023). Artificial intelligence for natural product drug discovery. *Nature Reviews Drug Discovery, 22*(11), 895–916.
20. Jasial, S., Hu, Y., & Bajorath, J. (2016). Determining the degree of promiscuity of extensively assayed compounds. *PLoS One, 11*(4), e0153873.
21. Hartenfeller, M., & Schneider, G. (2011). De novo drug design. *Methods in Molecular Biology, 672*, 299–323.
22. Fortin, F.-A., et al. (2012). DEAP: Evolutionary algorithms made easy. *The Journal of Machine Learning Research, 13*(1), 2171–2175.
23. Xu, Q.-Q., et al. (2023). Animal models of Alzheimer's disease: Preclinical insights and challenges. *Acta Materia Medica, 2*(2), 192–215.
24. Nicolaou, C. A., & Brown, N. (2013). Multi-objective optimization methods in drug design. *Drug Discovery Today: Technologies, 10*(3), e427–e435.
25. Trott, O., & Olson, A. J. (2010). AutoDock Vina: Improving the speed and accuracy of docking with a new scoring function, efficient optimization, and multithreading. *Journal of Computational Chemistry, 31*(2), 455–461.
26. Gillet, V. J. (2003). Application of evolutionary algorithms to combinatorial library design. In *Soft computing approaches in chemistry* (pp. 1–30). Springer.
27. DiMasi, J. A., Hansen, R. W., & Grabowski, H. G. (2003). The price of innovation: New estimates of drug development costs. *Journal of Health Economics, 22*(2), 151–185.
28. Lipinski, B., et al. (1997). Expectation maximization reconstruction of positron emission tomography images using anatomical magnetic resonance information. *IEEE Transactions on Medical Imaging, 16*(2), 129–136.
29. Rees, M. I., et al. (2006). Mutations in the gene encoding GlyT2 (SLC6A5) define a presynaptic component of human startle disease. *Nature Genetics, 38*(7), 801–806.
30. Rodríguez-Borges, C. G., & Sarmiento-Sera, A. (2011). Dimensionado mediante simulación de sistemas de energía solar fotovoltaica aplicados a la electrificación rural. *Ingeniería Mecánica, 14*(1), 13–21.
31. Hadka, D., & Reed, P. (2012). Diagnostic assessment of search controls and failure modes in many-objective evolutionary optimization. *Evolutionary Computation, 20*(3), 423–452.
32. Glen, R. C., & Payne, A. (1995). A genetic algorithm for the automated generation of molecules within constraints. *Journal of Computer-Aided Molecular Design, 9*(2), 181–202.
33. Landrum, G. A., Penzotti, J. E., & Putta, S. (2006). Feature-map vectors: A new class of informative descriptors for computational drug discovery. *Journal of Computer-Aided Molecular Design, 20*(12), 751–762.
34. Ramsundar, B., et al. (2019). *Deep learning for the life sciences: Applying deep learning to genomics, microscopy, drug discovery, and more*. O'Reilly Media.

Part III

Pharmacological Applications of AI in Phytopharmacology

AI Tools for Designing Phytopharmacological Agents

5

Muhammad Israr Khan and Gil Won Kim

Abstract

Artificial intelligence (AI) is revolutionizing phytopharmacology by addressing the complexity, diversity, and labor-intensive processes associated with plant-based drug discoveries. The traditional method of phytopharmacology utilizes intensive screening and disjointed data integration, making it difficult to reproduce the results and translate them into new applications. However, AI makes it easier to process big data sets, making it possible to quickly identify phytochemicals and to determine their potency, molecular targets, and safety profiles. Machine learning algorithms and deep learning provide assistance in identifying patterns from disorganized phytochemical databases, making it possible to rank new bioactive leads and understand their mode of action. Together with advancements in this area, new tools and computational models assisted by AI have been developed to aid pharmacological research. Omics-based databases integrate genomics, transcriptomics, proteomics, and metabolomics, thus provide assistance in understanding complex interactions between plants and their compounds. Whereas computing algorithms helps in model-based prediction of drug target interaction, pharmacokinetics, and toxicities. De novo molecular design tools helps in improving plant-based lead compounds to be safer, easier to absorb, and more efficacious while maintaining their plant-based sources. Instead of separate tools, novel AI frameworks work in uniting concepts of system biology, network pharmacology, and omics-based analysis into single predictive models to be used in phytopharmacology research. These frameworks allow understanding of phytochemical-based drugs that target multiple sites simultaneously, making it possible to understand plant drugs comprehensively. Though

M. I. Khan · G. W. Kim (✉)
Institute of Agriculture and Life Sciences, Gyeongsang National University, Jinju-si, South Korea
e-mail: gwkim@gnu.ac.kr

Y. Bibi et al. (eds.), *Artificial Intelligence in Phytochemistry*,
https://doi.org/10.1007/978-981-92-0875-3_5

many advancements have been achieved through AI and phytopharmacology, techniques such as interpretability of AI models, standardizing big data, and addressing ethical questions must be considered to make it relevant to human health. Despite these limitations, the union of AI and phytopharmacology has introduced a new philosophy known as prediction-based innovation in comparison to traditional description-based approaches associated with plant-based drugs. This chapter discusses progress in the application of AI tools such as generative chemistry and omics-based databases to comprehensively reform phytopharmacology-based innovation for safer, faster, and more personalized approaches to drug production.

Keywords

Molecular targets · Safety profiles · Machine learning · Deep learning · Bioactive leads ranking

5.1 Introduction

These days, the demand for herbal medicinal plants has been significantly increasing in pharmaceutical, food, cosmetics, and perfumery industries [1]. During COVID-19, plant-based medicines are proved to be some of the most effective methods for treating and preventing mild to moderate infections [2]. Recent technological advancements have also facilitated new approaches for natural product–based drug discovery, with bioinformatics and artificial intelligence strongly supporting research and development in this field. AI now enables rational phytochemical design by combining ethnobotanical knowledge with predictive analytics and virtual screening [3]. In drug discovery and development, artificial intelligence (AI) has become a disruptive force. AI utilizes computing power to analyze large volumes of data and generate insights that are difficult for human researchers to obtain independently. It encompasses a wide range of approaches and technologies, including NLP (natural language processing), deep learning, and machine learning, which have varying impacts on the pharmaceutical industry [2].

5.2 AI Platform for Plant-Based Drug Design

Artificial intelligence with machine learning and deep learning techniques is being increasingly applied to modern drug discovery and development. Machine learning (ML) allows a system to learn from data in an automated fashion with or without explicit programming. In supervised ML, with input data and desired outcomes, predictive models are built that help in disease diagnosis, predicting the efficacy of drugs, and their ADMET characteristics (*Absorption, Distribution, Metabolism, Excretion, and Toxicity*). Unsupervised ML, without predefined outcomes, analyzes data to discover new sub types of diseases and potential targets. These technologies

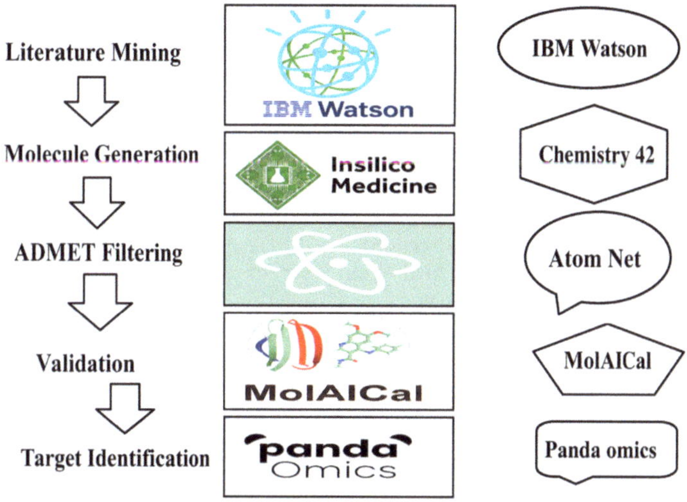

Fig. 5.1 AI-driven phyto drug design pipeline

have been extremely applicable to target identification and validation, drug design, drug repurposing, clinical trial decisions, and biomedical data analysis, along with the whole drug development process [4]. In the early stages of drug discovery, AI facilitates the creation of novel lead compounds that shows promising biological activities. Machine learning models help to identify and optimize target-specific molecules by predicting their safety, efficacy, and molecular interactions. By integrating fragment-based screening, computational modeling, and data-driven reasoning (Fig. 5.1), AI enhances the optimization of lead compounds. Ultimately, de novo compound designs and virtual screening approaches are replacing traditional biochemical assays for efficacy and toxicity testing [5]. This transformation allows drug developers to conduct efficient and cost-effective virtual experiments, reducing the time and resources required for laboratory-based drug evaluation [6].

5.2.1 IBM Watson for Drug Discovery (WDD) Platform

IBM WDD is an AI-driven next-generation platform that leverages the cognitive computing and natural language processing to analyze millions of pages of scientific literature and extract biological insights from them. It identifies and ranks potential genes and proteins through the interpretation of complex relationships among genes, diseases, and drugs [7]. Key features of WDD include the processing of unstructured text with contextual understanding in the field of biological interactions, presentations of these networks in an interactive visualization format, and the extraction of biological relationships (Fig. 5.2). Both rule-based and machine learning approaches of WDD are used to identify semantic relationships as gene-disease relationships between entities due to the co-occurrence of these terms within a

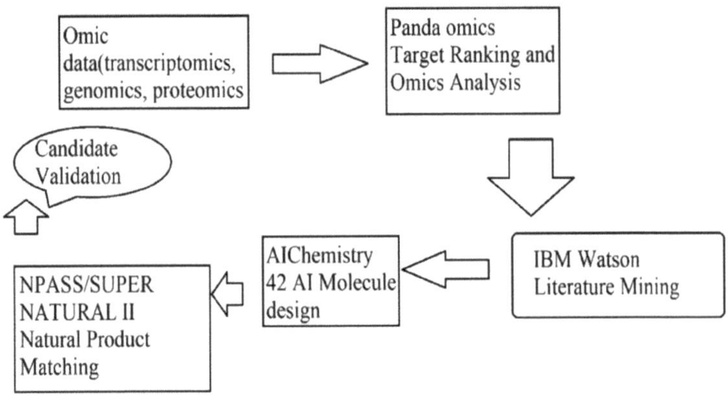

Fig. 5.2 Real-time AI thread for natural product discovery

sentence [5]. The tool is disease-agnostic, enabling the detection of shared gene–drug relationships across therapeutic areas. Integration of unstructured text data with structured sources, such as tables and figures, provides comprehensive insights into complex biological systems using WD [5].

The second major function of WDD is predictive analytics, which allows researchers to evaluate and rank lists of potential genes, diseases, or drugs. This function utilizes linguistic pattern recognition instead of relying on established biological knowledge to find novel relationships between genes and diseases that have not been explicitly mentioned in the literature [5]. Predictive analysis applies a list of known entities to build a conceptual space based on textual similarity, uncovering previously unappreciated connections. Its more specific applications includes the study of traumatic heterotopic ossification (tHO); WDD has been employed in two major ways: first, by extracting biological relationship networks from existing literature to uncover relevant patterns and associations related to tHO of various etiologies and second, by employing its predictive analytics model to identify new gene candidates potentially involved in its development.

5.2.2 Insilico Medicine

Deep learning (DL) has changed the face of drug discovery as it is now able to automatically identify patterns in complex, nonlinear data without resorting to any explicit feature extraction. It has been successfully applied in various fields such as image and speech recognition and has recently been adapted for small molecule design [8]. Insilico Medicine is among the first companies to use deep adversarial models to generate novel molecular compounds [9]. Since then, the number of DL frameworks that integrate generative models with reinforcement learning has increased in chemistry and pharmacology to design new molecules with desired pharmacological profiles [10]. These were followed by several de novo drug designs validated through in vitro and in vivo studies [11] Generative chemistry is one of the fastest-expanding fields in drug discovery today [12], with platforms like Chemistry42 driving innovation by combining AI, computational modeling, and medicinal chemistry expertise.

Chemistry42, which was introduced in 2020 by Insilico Medicine, merges advanced generative AI algorithms with medicinal and computational chemistry to design novel molecular entities in an efficient way. To date, the platform has been utilized by over 20 pharmaceutical companies. It operates through a secure, customizable workflow divided into three main stages: generation, learning, and analysis. In the generation phase, an ensemble of more than 40 generative models runs in parallel to create new molecular structures. These will be filtered and scored with respect to their properties and ADME profiles by using 2D and 3D modules. During analysis phase, molecules were graded on the basis of their synthetic accessibility, novelty, and diversity. Users can configure ligand-based or structure-based workflows, and they can upload ligand or protein data and modify reward modules, their thresholds, and diversity parameters. With an intuitive web interface and various auto-configuration attributes, the Chemistry42 platform represents a powerful and efficient tool for data-driven drug discovery [12].

5.2.3 Atomwise (ATOMNEt)

Atom Net is another major advancement in artificial intelligence, which has driven drug discovery as the first deep convolutional neural network (DCNN) designed for structure-based prediction of molecular binding affinity [13, 14]. In contrast with earlier ligand-based models, Atom Net incorporates the three-dimensional structural information of both ligands and their target binding sites, which thus enable them to learn complex biochemical interactions directly from spatial data. The DCNN architecture was initially developed for image and speech recognition [13]; it harnesses spatial locality, similar to biochemical interactions such as hydrogen bonding and π–π stacking [15]. This localized strategy enables Atom Net to effectively combine simple atomic interactions at a hierarchical level to form complex molecular binding patterns and thus increase prediction efficiency and performance [16]. Unlike conventional multi-task deep neural networks (MT-DNNs) that employ within ligand-based modeling, where Extended Connectivity Fingerprints (ECFPs) fingerprints and a significant existing dataset of assays are required, Atom Net enables predictions for new targets even in the absence of existing ligands. The fusion of both ligand and target structural knowledge addresses significant flaws within conventional approaches, including biasness of the data, lack of interpretability, and inability to model new interaction modes.

To evaluate its performance, Atom Net is benchmarked against commonly used datasets, including DUDE (Directory of Useful Decoys: Enhanced) and a ChEMBL-20 property-matched decoy (PMD) dataset [17]. The DUDE benchmark from the Shoichet Lab at UCSF includes 102 protein targets with 22,886 active compounds, each with 50 PMDs, designed to minimize analogue bias in their generation by scaffold-based clustering and selection [18]. Using this benchmark, 72 targets were used for training and 30 for testing in Atom Net. For further evaluation, a DUDE-like dataset was derived from version 20 of ChEMBL [19]. This version contained 123,102 total actives across 348 protein targets that had been filtered for high-confidence binding data, or $IC_{50}/K_i < 1$ μM, and structural resolution less than 2.5 Å [20]. Each active had been matched with 30 decoys sourced from the ZINC database, while scaffolds had been clustered using Bemis-Murcko methods to ensure chemical diversity. Here is a summary table you can insert on Page 6 after Paragraph 2 to provide a clear comparative overview of the AI platforms discussed.

5.2.4 MolAICal: A Computational Tool for 3D Structure-Based Drug Design Targeting Proteins

MolAICal is an advanced software package developed for three-dimensional (3D) drug design targeting protein binding pockets by integrating deep learning with classical computational algorithms. Traditional deep learning models can generate one-dimensional (1D) or two-dimensional (2D) molecular representations but struggle with accurate 3D ligand generation. MolAICal addresses this challenge through two Java-based modules, one combining a genetic algorithm with a deep learning model trained on FDA-approved drug fragments and the other integrating molecular docking with a deep learning model trained on drug-like compounds from the ZINC database. Unlike classical de novo design software such as LigBuilder or Open Growth, which require manual fragment selection, MolAICal autonomously generates valid, diverse, and drug-like fragments using deep generative models. It employs a Vinardo scoring system trained on the PDB bind database [22], offering superior binding assessment compared to Auto Dock Vina. [24] Furthermore, MolAICal eliminates the dependency on pre-existing ligand databases by generating sufficient numbers of drug-like ligands for virtual screening. Its Java-based design ensures cross-platform compatibility, enabling efficient operation on multi-core CPUs across Linux and Windows environments.

To illustrate its potential, MolAICal (open-source software platform for AI-driven drug design) is applied to two represent protein targets: the glucagon receptor, a membrane protein that regulates blood glucose, and SARS-CoV-2 main protease, which is essential for coronavirus replication. Compared to the COVID-19 Docking Server, a web-based platform based on SARS-CoV-2 developed with PHP [23], HTML, and JavaScript, the main difference between MolAICal is that it represents a standalone software program with an integrated deep learning approach. Whereas the COVID-19 Docking Server relies on tools like Open Babel [25], Auto Dock Vina [26], and Co Dock [23], MolAICal combines deep learning-based ligand generation with classical algorithms for the growth, optimization, and filtering of ligands. Experimental results showed that MolAICal was able to generate several ligands exhibiting both high and low 3D structural similarities to crystal ligands of GCGR and SARS-CoV-2 M pro. The tool and M-pro results are shared publicly on

GitHub to support global COVID-19 drug discovery efforts. Overall, MolAICal stands out as a robust, efficient, and freely available 3D drug design platform that closes the gap between deep learning and classical computational chemistry in rational drug development. The studied results of SARS-CoV-2 M pro are shared freely at https://github.com/MolAICal/COVID-19/tree/master/mpro to help scientists develop new potential drugs for COVID-19. MolAICal not only provides a strategy to solve the issue of 3D ligand generation in the protein pocket but also gives a reliable, free-of-charge, and effective software tool for rational drug design [27].

5.2.5 Benevolent AI

Serious concerns about the existential risks from advanced artificial intelligence motivated Eliezer Yudkowsky [28] and others to introduce the concept of "Friendly AI" or FAI [29, 30]. According to Yudkowsky [28], FAI is at once a research field and the AI system resulting from that research, focused on ensuring that advanced machine intelligence acts in ways beneficial to humans while not harmful to others. Unfortunately, the discourse on FAI has often been conducted in a vein of excessive fear, reducing objective evaluation. Typically, the supporters of the idea of FAI argue that since AI systems can differ fundamentally from humans, they will inevitably do so, and extreme caution is required to avoid selfish or harmful behavior. Further, they say that because AIs could be dangerous, it is safer to limit or control them. For example, Steve Omohundro [29] has suggested that AIs without explicit prosocial goals might behave like human sociopaths in seeking resources—a claim widely cited within the FAI community.

Critics argue, however, that such pessimistic assumptions do not take into account the given evolutionary and ethical mechanisms that intrinsically favor cooperation and altruism. Fox and Shulman [31] allude to several frameworks guaranteeing moral stability, including Trivers' theory of reciprocal altruism, Singer's expanding moral circle, and Wright's cooperative dynamics; they reject them as relevant only in cases of limited power differences. They also disregard the trend in human evolution toward increased moral capacity and reduced concentration on power differences. Their claim that optimally developed superintelligence could not possess the capacity for moral revision is contested by the fact that moral adaptation would already be included among its optimization priorities. Despite the evidence that more advanced intelligence can improve the chances of cooperation, rather than hostility, Fox and Shulman [30] continue to adopt a prudent attitude, warning that AGIs (Artificial General Intelligence) could be dangerous even without hostile intentions (Table 5.1), simply by using human resources in pursuits contradictory to human values. All in all, the tension between caution based on fear and the possibility of AI systems developing as cooperative and moral agents runs deep [32].

Table 5.1 Comparative summary of AI platforms for phyto drug design

Platform name	Core function	AI technique	Application stage	References
IBM Watson for Drug Discovery (WDD)	Literature mining and biological relationship extraction	Natural Language Processing (NLP), machine learning	Target identification and hypothesis generation	[5, 7]
Chemistry42	De novo molecular design and optimization	Generative models, reinforcement learning	Molecule generation and hit expansion	[12]
Atom Net	Structure-based prediction of molecular binding affinity	Deep Convolutional Neural Networks (DCNN)	ADMET filtering and binding prediction	[18, 21]
MolAICal	3D ligand generation and docking	Deep learning and genetic algorithms	Molecule generation and validation	[22, 23]

5.3 Integration of Natural Product Database

Computational methods have significantly advanced the identification and functional analysis of bioactive NPs, including in silico cheminformatics [33], molecular modeling and docking [33, 34], quantitative structure-activity relationship (QSAR) methods [33], and machine learning techniques [35, 36]. Integration of structural, quantitative activity, and species source data of NPs will enhance their effectiveness. Presently, more than one million NPs have been reported, with a sizeable portion experimentally interrogated for their biological activities [37]. General chemical databases and NP databases are the two most important resources in NP research. Commonly used chemical databases, including ChEMBL, PubChem Bio Assay, and Binding DB, have so far accumulated experimentally determined quantitative activity data [38–40]. The recommended general NP databases include Super Natural II (~325,000 NPs with 2D structures, physicochemical properties, and predicted targets) [41], UNPD (~229,000 NPs with 3D structures and species annotations), and ZINC (~80,000 NPs with 2D structures) [42]. Specialized NP databases focus on certain functional classes of NPs. Only ~4000 NPs have been reported with experimentally determined quantitative activity values from these databases, underlining the need for integrated and better-curated data (Fig. 5.2) [43].

5.3.1 NPASS (Neonatal Pain, Agitation, and Sedation Scale)

The NPASS database is accessible at https://bidd2.nus.edu.sg/NPASS/. It currently contains 35,032 unique natural products drawn from 25,041 species, which target 5863 molecular and cellular targets. It encompasses 288,002 NP-species and 222,092 NP-target associations [44] classified NPs into 18 chemical superclasses using the Classy Fire webserver, including alkaloids, benzenoids, lipids, lignin, and related compounds. Classes of compounds were provided by Djoumbou Feunang et al. [45]. The species source ranges from 6814 genera within Viridiplantae, Metazoan, fungi, and bacteria, which cover 67.8%, 9.4%, 7.9%, and 6.7%, respectively. Species entries with taxonomic information at the level of a species and genus are 59.6% and 93.1%, respectively. About 71% of NP entries are directly annotated from publications, while the rest are curated from existing databases.

The database targets include 2946 proteins, 121 protein family members, 143 protein complexes, 8 nucleic acids, 5 subcellular targets, 1352 microbial or pathogenic organisms, 1227 cell lines, and 51 tissue types [46]. Protein targets are further classified into enzymes, membrane receptors, ion channels, and transporters, with enzymes annotated according to the EC classification system. Targets of cell lines include 900 cancer lines, categorized by disease type, such as lung (161), brain (98), and leukemia (87) cancer lines. NPASS provides 446,552 quantitative activity records, including IC50/IC90/ID50 (23.2%), AC/AC40/AC50 (22.4%), MIC/MFC/MBC/FC (17.2%), GI/GI50/TGI (16.9%), percent inhibition (5.4%), EC50/EC90/ED50/ED90 (5.0%), and K_i (2.7%). Web access is enabled through HTML, PHP, and JavaScript, allowing browsing and searching by NP, target, or species source [43].

5.3.2 Super Natural II: A Database of Natural Products

Super Natural II is a free web-based database that enables easy access to natural products. Each entry includes detailed information on the chemical class of the NP, the predicted toxicity category of the NP, its (possible mechanisms of action) MoA, and the pathways involved. At the same time, the database provides toxicity alerts for single compounds; however, the absence of such alerts cannot be considered a guarantee of safety [47]. The toxicity prediction is carried out using the Pro Tox tool, which assesses structural similarities to known toxic compounds and performs well compared to other prediction methods [48]. So far, Super Natural II has been the first publicly available resource to combine such information; it incorporates approximately 326,000 unique compounds. The database consolidates MoA, pathway, and toxicity information in a single location. Its collection comes from 16 commercial suppliers and five publicly available databases.

5.3.3 Server, Database, and System Requirements

Super Natural II is based on a relational MySQL database (http://www.mysql.com). In order to handle the chemical information within the database, the MyChem package (http://mychem.sourceforge.net/) is used. For most of its functions, My Chem relies on the Open Babel toolbox (http://openbabel.org). The Super Natural II website is built using PHP (http://www.php.net/), JavaScript (http://www.java.com/), and Ajax; web access is enabled via an Apache HTTP Server (http://http.apache.org/). The site is best viewed in Mozilla Firefox and is also functional on alternative browsers like Google Chrome and Microsoft Internet Explorer.

5.4 Real-Time Applications and Integrated Threads

Real-time or near-real-time AI threads in modern drug discovery weave together various digital platforms to speed up the classically long process of target identification, molecular design, and validation of bioactivity. The IBM Watson platform, the successor of IBM Watson, serves as the enterprise cognitive engine for the retrieval and reasoning of biomedical data. It can mine large volumes of literature, clinical trial repositories, and omics records to give evidence-based insights in hours rather than weeks, hence driving dynamic hypothesis generation and prioritization of therapeutic targets in regulated environments [49]. Coupled to this knowledge layer, the Panda Omics platform from Insilco Medicine uses deep-learning models on multi-omics datasets for automatic ranking of disease-relevant genes, pathways, and biomarkers. It generates interpretable evidence reports and reduces the analytical turnaround time from raw data to target shortlists. Therefore, it is a core component of real-time target discovery [50]. Once the targets are prioritized, Chemistry42, another Insilico Medicine product, executes de novo molecular design by creating models with generative models that are constrained by physicochemical filters and synthetic accessibility scoring, producing chemically realistic scaffolds in days instead of months [51].

To improve biological relevance, these pipelines incorporate Super Natural II, a large natural-product database, and NPASS, a curated repository of quantitative natural-product activity and species-source data. Super Natural II offers over 325,000 structurally diverse compounds with annotated bioactivities and toxicity alerts that enable similarity searches and scaffold hopping of AI-designed hits [41]. Further refinement of this process is achieved with NPASS, providing experimentally measured activity values Including IC_{50}, K_i, EC_{50}, and MIC values to help validate or retrain in silico predictions of QSAR models in near real time [52, 53]. The common real-time discovery thread will therefore unfold as follows: raw transcriptomic data is fed into Panda Omics, yielding a ranked list of putative therapeutic targets; these are then mined by IBM Watson against the literature to distill a concise evidence graph in support of, or refuting, each candidate; Chemistry42 next generates candidate molecules optimized for the selected targets, applying rapid ADMET and synthetic feasibility filters; and finally, the proposed compounds are cross-referenced against NPASS and Super Natural II to identify natural analogues with experimentally validated activity profiles. This closed-loop thread transforms an initial biological signal into actionable chemical leads within several days, below common discovery timelines [51].

Beyond speed, these interrelated systems confer data provenance and interpretability: Watson provides explainable evidence reasoning; Panda Omics embeds target hypotheses in quantitative omics data; Chemistry42 ensures synthetic realism; and NPASS/Super Natural II databases enforce biological relevance. None of these platforms negates the need for experimental validation; laboratory assays remain necessary to confirm predictions made by computation. Collectively, they reflect a functional, real-world implementation of AI-driven, real-time drug discovery threads integrating knowledge retrieval, omics analytics, generative chemistry, and curated natural-product intelligence into a reproducible pipeline [51].

5.5 Conclusion

Overall, AI tools collectively support rational phytodrug design by integrating omics data, network pharmacology, predictive analytics, virtual screening, and generative chemistry into a unified discovery framework. Emerging approaches such as federated learning and explainable AI are expected to enhance data sharing, model transparency, and cross-platform reliability. However, the successful translation of AI-assisted phyto pharmacological innovations will depend on robust experimental validation and the development of adaptive regulatory frameworks to ensure safety, efficacy, and clinical.

References

1. Memarian, N., et al. (2011). Hierarchically assembled ZnO nanocrystallites for high-efficiency dye-sensitized solar cells. *Angewandte Chemie, International Edition, 50*(51), 12321.
2. Panda, A. K., et al. (2022). AYUSH-64: A potential therapeutic agent in COVID-19. *Journal of Ayurveda and Integrative Medicine, 13*(2), 100538.
3. Zang, J., et al. (2025). Early identification of stroke symptoms and risk factors using the BE FAST method: Benefits of early intervention in high-risk populations. *Frontiers in Neurology, 16*, 1630384.
4. Mak, K.-K., & Pichika, M. R. (2019). Artificial intelligence in drug development: Present status and future prospects. *Drug Discovery Today, 24*(3), 773–780.
5. Martin, R. L., et al. (2018). Hybrid natural language processing for high-performance patent and literature mining in IBM Watson for drug discovery. *IBM Journal of Research and Development, 62*(6), 8:1–8:12.
6. Lake, B. M., Linzen, T., & Baroni, M. (2019). Human few-shot learning of compositional instructions. *arXiv preprint*, arXiv:1901.04587. https://doi.org/10.48550/arXiv.1901.04587
7. Bakkar, N., et al. (2018). Artificial intelligence in neurodegenerative disease research: Use of IBM Watson to identify additional RNA-binding proteins altered in amyotrophic lateral sclerosis. *Acta Neuropathologica, 135*(2), 227–247.
8. Chen, H., et al. (2018). The rise of deep learning in drug discovery. *Drug Discovery Today, 23*(6), 1241–1250.
9. Kadurin, A., et al. (2016). The cornucopia of meaningful leads: Applying deep adversarial autoencoders for new molecule development in oncology. *Oncotarget, 8*(7), 10883.
10. Vanhaelen, Q., Lin, Y.-C., & Zhavoronkov, A. (2020). The advent of generative chemistry. *ACS Medicinal Chemistry Letters, 11*(8), 1496–1505.
11. Zhavoronkov, A., et al. (2019). Deep learning enables rapid identification of potent DDR1 kinase inhibitors. *Nature Biotechnology, 37*(9), 1038–1040.

12. Zhavoronkov, A., Kamya, P., & Pun, F. (2022). *From paper to industrial-scale platform: A 3-year behind the paper journey from GENTRL to Chemistry42* (Bioengeering & biotechnology). Springer Nature. http://bioengineeringcommunity.nature.com/posts/from-paper-to-industrial-scale-platform-a-3-year-behind-the-paper-journey-from-gentrl-to-chemistry42. Accessed 21 Sept 2022.
13. LeCun, Y., Bengio, Y., & Hinton, G. (2015). Deep learning. *Nature, 521*(7553), 436–444.
14. Krizhevsky, A., Sutskever, I., & Hinton, G. E. (2012). ImageNet classification with deep convolutional neural networks. *Advances in Neural Information Processing Systems, 6*(60), 84–90.
15. Yosinski, J., et al. (2015). Understanding neural networks through deep visualization. *arXiv preprint*, arXiv:1506.06579. https://doi.org/10.48550/arXiv.1506.06579
16. Ramsundar, B., et al. (2015). Massively multitask networks for drug discovery. *arXiv preprint*, arXiv:1502.02072. https://doi.org/10.48550/arXiv.1502.02072
17. Wang, Y., et al. (2012). PubChem's BioAssay database. *Nucleic Acids Research, 40*(D1), D400–D412.
18. Mysinger, M. M., et al. (2012). Directory of useful decoys, enhanced (DUD-E): Better ligands and decoys for better benchmarking. *Journal of Medicinal Chemistry, 55*(14), 6582–6594.
19. Bento, A. P., et al. (2014). The ChEMBL bioactivity database: An update. *Nucleic Acids Research, 42*(D1), D1083–D1090.
20. Desaphy, J., et al. (2015). sc-PDB: A 3D-database of ligandable binding sites—10 years on. *Nucleic Acids Research, 43*(D1), D399–D404.
21. Dahl, G. B., Kostøl, A. R., & Mogstad, M. (2014). Family welfare cultures. *The Quarterly Journal of Economics, 129*(4), 1711–1752.
22. Wang, R., et al. (2004). The PDBbind database: Collection of binding affinities for protein–ligand complexes with known three-dimensional structures. *Journal of Medicinal Chemistry, 47*(12), 2977–2980.
23. Kong, R., et al. (2020). COVID-19 Docking Server: A meta server for docking small molecules, peptides and antibodies against potential targets of COVID-19. *Bioinformatics, 36*(20), 5109–5111.
24. Wang, Z., et al. (2016). Comprehensive evaluation of ten docking programs on a diverse set of protein–ligand complexes: The prediction accuracy of sampling power and scoring power. *Physical Chemistry Chemical Physics, 18*(18), 12964–12975.
25. O'Boyle, N. M., et al. (2011). Open Babel: An open chemical toolbox. *Journal of Cheminformatics, 3*(1), 33.
26. Trott, O., & Olson, A. J. (2010). AutoDock Vina: Improving the speed and accuracy of docking with a new scoring function, efficient optimization, and multithreading. *Journal of Computational Chemistry, 31*(2), 455–461.
27. Bai, Q., et al. (2020). MolAICal: A software tool for 3D drug design of protein targets by artificial intelligence and classical algorithm. *Briefings in Bioinformatics, 22*(3), bbaa161.
28. Yudkowsky, E. (2008). Artificial intelligence as a positive and negative factor in global risk. In N. Bostrom & M. M. Ćirković (Eds.), *Global catastrophic risks*. Oxford University Press.
29. Omohundro, S. (2008). The basic ai drives-proceedings of the first agi conference. In *Frontiers in artificial intelligence and applications*. IOS Press.
30. Fox, J., & Shulman, C. (2010). Superintelligence does not imply benevolence. In K. Mainzer (Ed.), *ECAP10: VIII European conference on computing and philosophy* (Vol. 10, pp. 456–462). Verlag Dr. Hut.
31. Sotala, K. (2010). From mostly harmless to civilization-threatening: Pathways to dangerous artificial general intelligences. In K. Mainzer (Ed.), *ECAP10: VIII European conference on computing and philosophy* (Vol. 10, pp. 443–450). Verlag Dr. Hut.
32. Waser, M. (2011). Rational universal benevolence: Simpler, safer, and wiser than "friendly AI". In *International conference on artificial general intelligence*. Springer.
33. Rodrigues, T., et al. (2016). Counting on natural products for drug design. *Nature Chemistry, 8*(6), 531–541.

34. Chen, X., Ung, C. Y., & Chen, Y. (2003). Can an in silico drug-target search method be used to probe potential mechanisms of medicinal plant ingredients? *Natural Product Reports, 20*(4), 432–444.
35. Rupp, M., et al. (2014). Machine learning estimates of natural product conformational energies. *PLoS Computational Biology, 10*(1), e1003400.
36. Ung, C. Y., et al. (2007). Are herb-pairs of traditional Chinese medicine distinguishable from others? Pattern analysis and artificial intelligence classification study of traditionally defined herbal properties. *Journal of Ethnopharmacology, 111*(2), 371–377.
37. Demain, A. L., & Vaishnav, P. (2011). Natural products for cancer chemotherapy. *Microbial Biotechnology, 4*(6), 687–699.
38. Gaulton, A., et al. (2017). The ChEMBL database in 2017. *Nucleic Acids Research, 45*(D1), D945–D954.
39. Wang, J., et al. (2005). Traditional Chinese medicine information database. *Clinical Pharmacology & Therapeutics, 78*(1), 92–93.
40. Gilson, M. K., et al. (2016). BindingDB in 2015: A public database for medicinal chemistry, computational chemistry and systems pharmacology. *Nucleic Acids Research, 44*(D1), D1045–D1053.
41. Banerjee, P., et al. (2015). Super natural II—A database of natural products. *Nucleic Acids Research, 43*(D1), D935–D939.
42. Sterling, T., & Irwin, J. J. (2015). ZINC 15–ligand discovery for everyone. *Journal of Chemical Information and Modeling, 55*(11), 2324–2337.
43. Zeng, X., et al. (2017). NPASS: Natural product activity and species source database for natural product research, discovery and tool development. *Nucleic Acids Research, 46*(D1), D1217–D1222.
44. Zhang, C., et al. (2017). Pharmacological relationships and ligand discovery of G protein-coupled receptors revealed by simultaneous ligand and receptor clustering. *Journal of Molecular Graphics and Modelling, 76*, 136–142.
45. Djoumbou Feunang, Y., et al. (2016). ClassyFire: Automated chemical classification with a comprehensive, computable taxonomy. *Journal of Cheminformatics, 8*(1), 61.
46. Xue, R., et al. (2012). TCMID: Traditional Chinese medicine integrative database for herb molecular mechanism analysis. *Nucleic Acids Research, 41*(D1), D1089–D1095.
47. Banerjee, P., et al. (2014). Super natural II—A database of natural products. *Nucleic Acids Research, 43*(D1), D935–D939.
48. Drwal, M. N., et al. (2014). ProTox: A web server for the in silico prediction of rodent oral toxicity. *Nucleic Acids Research, 42*(W1), W53–W58.
49. SHekhar, A., Gupta, R., & Sharma, S. K. (2025). IBM Watson health growth strategy: Is artificial intelligence (AI) the answer. *Communications of the Association for Information Systems, 57*(1), 63.
50. Kamya, P., et al. (2024). PandaOmics: An AI-driven platform for therapeutic target and biomarker discovery. *Journal of Chemical Information and Modeling, 64*(10), 3961–3969.
51. Ivanenkov, Y. A., et al. (2023). Chemistry42: An AI-driven platform for molecular design and optimization. *Journal of Chemical Information and Modeling, 63*(3), 695–701.
52. Zeng, X., et al. (2018). NPASS: Natural product activity and species source database for natural product research, discovery and tool development. *Nucleic Acids Research, 46*(D1), D1217–D1222.
53. Zhao, H., et al. (2023). NPASS database update 2023: Quantitative natural product activity and species source database for biomedical research. *Nucleic Acids Research, 51*(D1), D621–D628.

Predicting Efficacy and Toxicity of Phytochemical Drugs with AI Models

6

Ambreen Najaf and Muhammad Umair Asghar

Abstract

Attention to phytochemicals as beneficial agents is increasing, and several efforts have been made to assess their worth and well being using several computational methods. Traditional approaches to drug development have contributed significantly to pharmacological advancement but often involve high investment, are cumbersome, and suffer from poor predictive accuracy. Recent advances in AI now permit data-driven prediction of the pharmacological properties of natural products more efficiently and precisely. This chapter describes the role of artificial intelligence models in predicting the efficacy-toxicity profile of phytochemical drugs by using models such as machine learning, deep learning, and neural networks. Additionally, it discusses the application of AI in identifying bio active compounds, performing in silico analysis of molecular interaction, and enhancing drug-likeness with minimal adverse properties. A strategy that incorporates systems biology, molecular docking, and multi-omics data for improving predictive accuracy is also discussed. In this chapter, the emerging role of explainable AI in model interpretability and transparency is explored. Despite such advances, challenges persist regarding dataset's limitations, model vulnerability, and standardized computational frameworks. Overall, this chapter highlights how AI provides predictive, scalable, and personalized phytochemical drug development by offering a game-changing platform for safer and more efficient natural product-based therapeutics.

A. Najaf (✉)
Department of Botany, Rawalpindi Women University, Rawalpindi, Pakistan

M. U. Asghar
Department of Animal Nutrition and Feed Science, Faculty of Biology and Animal Science, Wrocław University of Environmental and Life Sciences, Wrocław, Poland

Y. Bibi et al. (eds.), *Artificial Intelligence in Phytochemistry*,
https://doi.org/10.1007/978-981-92-0875-3_6

Keywords

Systems biology · Molecular docking · Multi-omics data integration · Explainable AI

6.1 Introduction

It has long been recognized that phytochemicals are naturally occurring bioactive products produced from plants, a crucial basis of therapeutic agents for health in humans [1]. However, traditional drug development from phytochemicals is yet an expensive, time-consuming, and experimentally challenging process, although they possess very promising pharmacological potential-from antioxidant to anticancer and anti-inflammatory effects [22]. As the majority of candidate drugs fail during preclinical or clinical tests because of safety or efficacy problems, early-phase efficacy and toxicity assessment becomes extremely important [33]. Such limitations have been overcome by AI and machine learning techniques, which, in contrast to the experimental approaches of the past, are more precise and efficient in dealing with complex data sets and detecting molecular patterns, which allow for predicting biological events [32].

Current progress within AI-driven models, such as Deep Learning, Graph Neural Networks, and Support Vector Machines, also transformed this field of computational pharmacology by empowering the direct forecast of pharmacological and toxicological results from chemical constructions. Some of the models allow for rapid screening of natural substances with respect to their potential therapeutic efficacy in the context of phytochemicals while concurrently considering possible side effects.

AI systems, such as DeepChem, ProTox-II, and ADMETlab, use chemical descriptors, molecular fingerprints, and biological data to predict the absorption, distribution, metabolism, excretion, and toxicity profiles of phytochemicals. These computational insights accelerate the translation of bioactive natural compounds into clinical testing and greatly reduce the need for animal testing (Table 6.1).

Moreover, a systems-level comprehension of the mechanism of action of phytochemicals is enabled by the incorporation of multi-omics data-integrating AI-based frameworks: genomic, transcriptomic, proteomic, and metabolomic data on the mode of action of phytochemicals, according to [39]. Issues that include but are not limited to the "black-box" nature of models using deep learning, lack of standardized datasets, and restricted structural variety in chemical libraries that are currently accessible get in the way of the reliability and interpretability of predictions, according to [23]. Explainable AI approaches are becoming increasingly important in the field of phytochemical research due to a growing need for AI models with high biological interpretability and transparency, for example [7].

In any case, it follows that phytochemical drug discovery combined with AI technology represents a paradigm shift in pharmaceutical research. AI-driven approaches hold great potential for improved drug design by reducing safety risks and accelerating the discovery of natural medicinal molecules (Fig. 6.1). This will be achieved by the ability to accurately predict efficacy and toxicity profiles, as stated by [26]. This

Table 6.1 Datasets and computational resources

Category	Name/tool	Description and use in AI modeling	References
Phytochemical databases	PubChem	Comprehensive chemical repository with structural, physicochemical, and biological data; used for generating molecular fingerprints and features for AI models	[20]
	ChEMBL	Curated bioassay data including IC_{50}, Ki, and potency; supports QSAR and efficacy modeling	[2]
	NPASS	Links natural products with their biological sources and experimentally validated activities for AI-based efficacy prediction	[37]
	IMPPAT	l Indian medicinal plants database providing phytochemistry and therapeutic information for ethnopharmacological AI studies	[21]
	TCMID	Traditional Chinese Medicine Integrated Database combining phytochemical data with traditional therapeutic uses	[19]
Toxicity databases	Tox21	High-throughput screening dataset for identifying potential toxic effects of chemicals, including phytochemicals	[14]
	ToxCast	Contains biological activity profiles for thousands of compounds, supporting ML-based toxicity predictions	[14]
	CTD (Comparative Toxicogenomics Database)	Links chemical exposures with gene and disease data to assist AI models in toxicity assessment	[10]
Computational frameworks	DeepChem	Open-source Python library for deep learning in drug discovery, QSAR, and toxicity prediction	[25]
	Scikit-learn/ TensorFlow/ PyTorch	Core machine learning and deep learning frameworks used to build, train, and validate predictive AI models	[24]
	AutoDock/ SwissDock/MOE/ Schrödinger Maestro	Docking and molecular simulation tools integrated with AI models to predict ligand–target binding affinities	[38]
ADMET prediction tools	SwissADME	Web-based platform to assess drug-likeness and pharmacokinetics using ML models	[36]
	pkCSM/ ADMETlab 2.0	Predicts ADMET properties and toxicity using graph-based and neural network models	[36]

(continued)

Table 6.1 (continued)

Category	Name/tool	Description and use in AI modeling	References
Computational infrastructure	Google Colab/ AWS/Microsoft Azure ML	Cloud-based AI platforms providing scalable computing power for deep learning on phytochemical datasets	[6]
	Apache Spark/ Dask	Distributed data-processing systems enabling large-scale parallel computation for molecular datasets	[33]
Data management principles	FAIR Data Standards	Framework ensuring data are Findable, Accessible, Interoperable, and Reusable, crucial for reproducible AI research	[34]

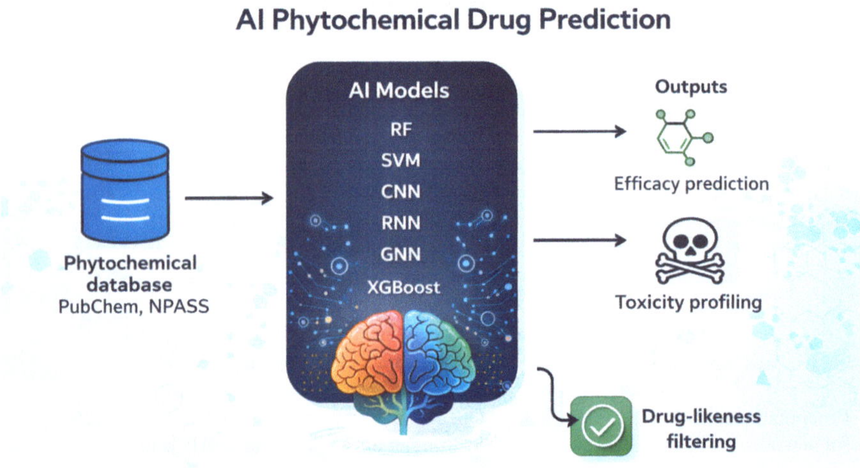

Fig. 6.1 AI workflow for phytochemical drugs prediction

review outlines the latest progress on recent AI model applications in phytochemical drug discovery. Emphasis is placed on methodology, databases, obstacles, and future approaches in developing natural drugs with precision and safety.

6.2 Artificial Intelligence in Modern Drug Discovery

"The increasing use of computationally intensive strategies to mine very large biological datasets for evidence of interactions between drugs and their bioactivity signatures a process which enables an accurate prediction of both potential efficacy and potential toxicities" all point to a new role for artificial intelligence in pharmaceuticals: a pivotal role in pharmaceutical discovery. Traditional drug discovery pipelines are very costly and take a long time to develop, which means they are predisposed toward a high failure rate because only a few candidate compounds are transferred from preclinical testing into market approval. Artificial intelligence, in particular its subfields machine learning and deep learning, provides data-driven methods for overcoming many of these challenges through automated molecular screening, lead optimization, and toxicity assessment. Integration of AI models with cheminformatics and bioinformatics identifies bioactive compounds, including natural product phytochemicals, at significantly higher speed and efficiency than traditional experimental methodologies [27].

AI algorithms in drug development use molecular descriptors and structural fingerprints to predict physico-chemical, pharmacokinetic, and pharmacodynamic properties of substances. Deep learning architectures using CNN and GNN can learn hierarchical molecular features automatically from chemical graphs and SMILES strings, while machine learning methods such as RF, SVMs, and GBMs have been used to model drug-target interactions. These not only improve the accuracy of prediction but also enable the virtual screening of thousands of phytochemicals for the identification of leads for therapeutic development.

In the last years, remarkable development of the AI-based DeepChem, AlphaFold, and ADMETlab integrated approaches has been seen to implement different computational frameworks for predicting the mode of drug action and molecular conformations [40]. Thus, AlphaFold, developed by DeepMind, revolutionized structural biology by predicting protein structures at almost experimental accuracy, thereby allowing more precise investigations on drug-target docking [13]. The AI-assisted ADMET models predict the behavior that may be assumed by candidate drugs within the biological system, hence facilitating early safety profile evaluation while reducing the demand for in vivo testing [5].

Now, AI presents a new paradigm in phytochemical drug discovery to overcome long-standing challenges related to the complexity of plant metabolomes and the diversity of chemical scaffolds. AI-driven cheminformatics might finally allow scientists to decipher at a rapid pace the correlation between the chemical structure and its biological function as a way to identify compounds with specific pharmacological properties. On the other side, the combination of multi-omics data, including transcriptomics, metabolomics, and genomics, into the AI workflow is now allowing systems-level awareness into how phytochemicals interrelate with biological ways.

With these developments, the full potential of AI in drug discovery has yet to be realized. Some of the challenges that are constant include high-quality and standardized datasets, the interpretability of complicated models, and transferability of predictions across many compound classes [23]. The synergy between AI and drug discovery is a face of modern pharmacology in evolution, where a more predictive, effective, data-driven approach is being offered for finding safe and effective therapeutic candidates, including those sourced from natural products like phytochemicals [32].

6.3 Phytochemical-Based Drug Discovery: Opportunities and Challenges

The great structural variety and high level of biological activity of natural chemicals have made phytochemical-based drug discovery feasible in modern pharmacology. Phytochemicals from medicinal plants are conventionally useful sources for new therapeutic drugs, providing the opportunity to modify their structural templates synthetically. Numerous natural compounds, including alkaloids, terpenoids, flavonoids, and saponins, exhibit a wide range of pharmacological activities, such as anticancer, antibacterial, anti-inflammatory, and neuroprotective properties. Screening phytochemicals is thus an excellent opportunity to discover lead compounds with low toxicity and high efficacy and will, in turn, help in developing safer and potent medication formulations.

In spite of this medicinal promise, there are several obstacles to medication discovery based on phytochemicals. Among the major drawbacks, the chemical composition of plant extracts is very complicated, which causes all efforts toward the separation of compounds, standardization, and repeatability to become tougher to achieve [19]. Besides, for most phytochemicals, pharmacokinetics and bioavailability are incompletely understood, which often leads to variable efficacy upon clinical applications [33]. Another critical barrier is the lack of thorough toxicity data because traditional screening approaches require much time and resources. These obstacles mean that it is hard for many promising phytochemicals to advance from laboratory research into the development of clinical medicine [11].

Recent progress within the field of AI and computational technologies can surmount these challenges. For example, virtual screening, investigation of SAR, and prediction of toxicity will be done for large libraries of phytochemicals using AI-based predictive models [35]. Coupled with AI approaches, phytochemical databases and bioinformatics tools will facilitate rapid identification of bioactive compounds with the purpose of optimization of their pharmacological properties toward a successful clinical outcome [28]. Thus, integration of traditional knowledge about medicinal plants with advanced AI-based predictive analytics holds great promise to surmount the existing challenges in phytochemical drug development.

6.4 AI Models for Predicting Efficacy of Phytochemicals

The data-driven modeling, AI has now become a game-changing tool in phytochemical research; it allows the forecasting of the biological use of the chemical obtained from plants. Experimentally screening phytochemicals for their medicinal potential had been an expensive and time-consuming job earlier. Researchers can now study huge chemical and biological data for the identification of complex structure activity relationships and pharmacological features based on AI-based computer models [41].

Scientists also observe that through the use of chemical fingerprints, molecular descriptors, and bioassay data, AI models can predict phytochemical efficacy in silico prior to experimental confirmation. This, therefore, presents an easy way of finding natural substances featuring antibacterial, antioxidant, and anticancer properties. AI, through the use of learning algorithms, therefore, enables high-throughput virtual screening and prioritizes promising bioactive compounds from hundreds of phytochemicals hosted in plant databases [19].

Various ML and AI approaches have been employed for this purpose. These include the most popular, such as SVMs, RFs, ANNs, and DL, models [4]. These models are trained on curated datasets with pre-identified chemical structures and the corresponding biological activities. The learning algorithms thus establish quantitative relationships between the chemical structure and activity; therefore, it is possible to predict the efficacy of new or untested phytochemicals [29].

Accuracy in efficacy prediction has been further improved by the effective integration of AI-driven QSAR models with hybrid deep learning systems integrated into molecular docking and molecular dynamics simulations [38]. A hybrid deep learning architecture, for example, has been able to find plant-based inhibitors against cancer-related enzymes that integrate the docking scores with physicochemical characteristics [12]. Such integrated models predict the potential pharmacological effects of candidate phytochemicals besides their binding affinity. AI-based efficacy prediction models have difficulties with data scarcity, model interpretability, and generalizability despite their encouraging results. Training and validation procedures are hampered by the lack of high-quality, consistent phytochemical datasets. Furthermore, even though deep learning models are reliable, they frequently function as "black boxes," making it challenging to understand why a particular compound is projected to be effective. In order to increase the transparency and reliability of phytochemical predictions, explainable AI (XAI) frameworks and hybrid modeling techniques are being developed [24].

In general, the use of AI in phytochemical efficacy prediction represents a paradigm change from conventional trial-based discovery to computational pharmacognosy, where predictive modeling and virtual screening speed up drug development and lower experimental costs. Zhou et al. [41] highlight that the future of phytochemical medication discovery and efficacy analysis will be greatly influenced by the ongoing development of AI algorithms in conjunction with expanding natural compound databases (Table 6.2).

6.5 Integration of Multi-omics and Systems Biology with AI

The prediction of phytochemical efficacy and toxicity is influenced by the integration of AI with systems biology and multi-omics data. In genomics and transcriptomics, deep neural networks and support vector machines are major AI techniques deployed in the field that allow for the identification of compound-gene interaction and, thus, prediction of the modulation exerted by phytochemicals on gene expression networks. These models allow for target discovery and explain how plant-derived compounds modulate genetic pathways associated with diseases. On the other hand, integration with proteomics is facilitated by convolutional neural

Table 6.2 Comparative overview of AI models used in phytochemical efficacy and toxicity prediction

Model type	Strengths	Use cases in phytochemical prediction	References
Random Forest (RF)	High accuracy with small–medium datasets; robust to noise; handles nonlinear relationships; interpretable feature importance	Toxicity prediction, bioactivity classification, ADMET property prediction of phytochemicals	[8]
Support Vector Machine (SVM)	Effective in high-dimensional spaces; works well with limited data; strong generalization	Classification of active vs inactive phytochemicals; toxicity and anticancer activity prediction	[18]
Convolutional Neural Networks (CNN)	Automatic feature extraction; strong performance on image-like and spatial data	lMolecular fingerprint analysis; structure–activity relationship (SAR) modeling; compound screening	[30]
Recurrent Neural Networks (RNN)	Captures sequential dependencies; suitable for time-series and sequence data	SMILES-based phytochemical modeling; prediction of molecular properties from chemical sequences	[3]
Graph Neural Networks (GNN)	Directly models molecular graphs; captures atomic interactions and topology	Prediction of binding affinity, toxicity, and bioactivity using molecular graph representations	[9]
Extreme Gradient Boosting (XGBoost)	High predictive performance; efficient; handles missing data well	ADMET prediction; multitask learning for efficacy and toxicity screening of phytochemicals	[6]

networks and graph neural networks, analyzing protein–ligand interactions and structure–function relationships that provide predictive information regarding the efficacy and off-target effects of phytochemicals (Fig. 6.2).

The objective of the metabolomics dimension is to understand how phytochemicals affect metabolic pathways and biochemical responses of an organism. Examples of machine learning approaches applied in finding metabolic biomarkers associated with hazardous or therapeutic effects of phytochemicals include PCA, random forests, and GBMs. AI-driven approaches allow modeling of multi-target and multi-pathway mechanisms in systems pharmacology and network biology, exposing antagonistic or synergistic interactions between bioactive substances. Graph-based learning and network pharmacology frameworks are very helpful in mapping such complex biological interactions.

Besides, AI algorithms supporting multi-omics data fusion enable the integration of various biological datasets to discover hidden patterns that interconnect genes, proteins, and metabolites [41]. These models, when applied together with systems biology, help in deciphering the phytochemical mechanisms of action using XAI

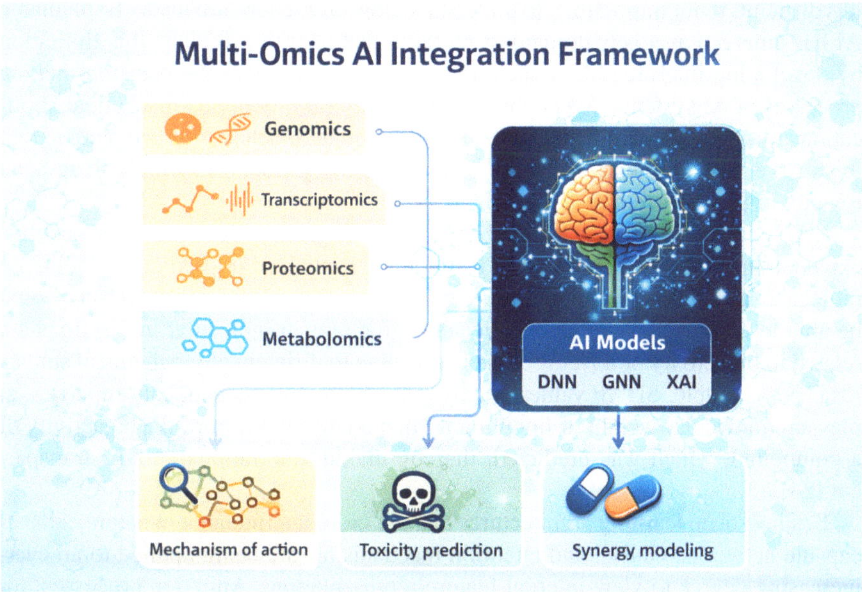

Fig. 6.2 Multi-omics integration framework

and pathway enrichment techniques. The interpretability allows researchers to connect computational predictions with experimental validation [24].

These integrated AI and systems biology frameworks also contribute to toxicity and synergy modeling apart from efficacy prediction by anticipating the combinatorial effects of phytochemicals and detecting possible hazardous interactions prior to clinical evaluation. Reinforcement learning and ensemble learning techniques have been used to enhance compound safety profiles and predict system-level toxicity responses [17]. Cloud-based AI platforms and bioinformatics tools such as TensorFlow, PyTorch, DeepChem, and Cytoscape have greatly improved the scalability and precision of such computational models [24].

In the near future, the integration of AI with systems biology and multi-omics will revolutionize precision phytomedicine by enabling personalized predictions of patient responses to plant-based treatments. Emerging technologies in knowledge graph-based modeling and federated learning by [41] will further enhance the interpretability and predictive power of AI-driven phytochemical research.

6.6 Explainable AI (XAI) and Model Interpretability

While the rapid acceptance of AI in phytochemical drug development has been accompanied by exciting gains in predictive accuracy, this rapid development has also brought increasing concerns over interpretability and transparency. Traditional deep learning models are highly accurate but often consist of "black boxes" where it is difficult, if not impossible, to understand how predictions are made. Explainable AI has emerged as a new domain of research that tries to overcome this limitation by developing interpretable models that provide insight into the decision-making processes of AI systems. XAI enhances trust and usability in pharmaceutical applications through the identification of which molecular, biochemical, and genetic features have most influence on model outputs in the prediction of the efficacy and toxicity of phytochemicals.

XAI techniques, such as LIME, SHAP, and Integrated Gradients, offer post hoc explanations with the attribution of every feature involved in prediction, including molecular descriptors, binding affinities, or toxic alerts [16]. These methods would be able to identify how some chemical substructures or molecular interactions are related to bioactivity or adverse effects according to different phytochemical studies [31]. For example, SHAP values may reveal which physicochemical properties, at play, are molecular weight or lipophilicity that contributes toward the bioactivity of a compound, which was predicted, thereby making the rationale of AI transparent [24].

Besides, deep learning architectures that are more interpretable in nature, such as capsule networks, GNNs, and attention mechanisms, are being applied to an ever-increasing degree to view internal feature representations. Attention processes, for example, will indicate which atoms or bonds in the molecular graph have the greatest influence on biological activity in phytochemical efficacy predictions. These are important understandings in designing novel phytochemicals or modifying already existing ones for improved therapeutic effects with low toxicity.

In the future, precision phytomedicine will be further enabled by integrations between explainable AI and knowledge graphs, causal inference models, and multi-omics data integration for further enhancements in interpretability of complex phytochemical-target interactions [38]. Conclusively, XAI represents a critical transition from black-box prediction models toward understandable frameworks by scientists, with the guarantee that AI-driven discoveries in phytochemical research remain reliable, interpretable, and physiologically relevant.

6.7 Limitations and Current Challenges

Although computational modeling and AI have significantly advanced phytochemical drug development, a number of factors still limit the interpretability, accuracy, and generalizability of the models. The major deficiencies concern the quality and diversity of the available datasets. Most of the phytochemical datasets are incomplete, inconsistent, or biased toward particular classes of compounds or target types and thus limit the ability of AI models to generalize across a variety of biological systems [10].

Another key challenge for phytochemical efficacy and toxicity prediction is interpretability or transparency of the complex deep learning models active. Despite their very high precision, those models act mostly like "black boxes," giving insight neither into the biological methods underlying their predictions nor the building blocks of their model. This idea has been established by [15]. Because of such opacity, pharmacologists and regulatory agencies are less believing, and this also hampers scientific validation. Incorporation of XAI frameworks in phytochemical modeling is just at its inception, despite their emergence as viable clarifications. The works of [40] have moved upon this idea.

Due to the differences in volume, complexity, and experimental platform, multi-omics data integration is even more challenging. The integration of genomics, transcriptomics, proteomics, and metabolomics requires sophisticated techniques of normalization and data fusion, while existing AI algorithms often cannot achieve a balance between computation complexity and biological relevance [19]. Besides, there is an extreme lack of in vitro and in vivo validation to verify AI-based predictions. Without empirical validation, the conclusion at the computation level remains theoretical, especially for new phytochemicals or those that are hardly studied. Uniform benchmarking procedures are also lacking across different research teams, which further constrains comparative analysis at the model performance level [38].

Data imbalance is still a serious technical problem in many cases, and even for the toxicity prediction jobs, either positive or negative samples are often underrepresented. Such imbalance may lead to biased models that overstate either efficacy or safety, a potential danger leading to further development of drugs. Another parallel situation involves small labs or institutions, particularly those from underdeveloped nations, being unable to apply these AI approaches due to limited computational resources.

Other evolving challenges alarm ethical and regulatory matters. In the lack of any legal contexts about AI-driven phytomedicine, standardization of validation processes and valuation of data privacy and justice in AI predictions remain very difficult to perform [24]. The limited variety within the training datasets, often from particular locations or species, raises a number of worries about model bias and ecological representation. In order to overcome such constraints, international collaboration will be required, developing and implementing transparent AI frameworks, highlighting interpretability and repeatability, and providing open-access phytochemical funds with standardized workflows for justification.

6.8 Future Perspectives

AI can bring a phytochemical drug development revolution, transforming a previously non-data-heavy pharmacological research field into an immensely data-driven predictive one. As deep learning, systems biology, and computational chemistry continuously evolve, AI models will be far more accurate in the prediction of phytochemical toxicity and efficacy. Combining multi-omics, such as proteomics, metabolomics, and genomes, will allow understanding of the interaction of phytochemicals in complex biological systems in detail to result in more narrow and successful therapeutic breakthroughs.

These are just a few of the new technologies that will enhance the accuracy of models and increase the efficiency of processing even more: federated learning, simulation of digital twins, and quantum computing. Such developments are capable of opening up quicker and more accurate in silico screening of phytochemicals by modeling interactions on the scale that was never previously conceivable. The interpretability, transparency, and confidence of XAI would be needed since regulators must accept and translate AI-driven predictions of phytochemical-based medication into the clinic.

The increased use of AI in research on natural products presupposes the use of open-source solutions and global databases in which researchers might share high-quality and standardized data, which in turn would lead to the mitigation of the current limitation due to the lack of data and its heterogeneity. Moreover, the latter notion is applied to the personalized phytomedicine, premised on the genetic and metabolic profile of the individual, which would also involve tailored therapeutic remedies with the minimum toxicity and the highest possible efficacy.

The sea change of the drug discovery procedure is most likely to be created in the future by AI-enhanced biotechnology and pharmacognosy. The near future studies will accelerate the creation of new phytochemicals that use biological knowledge with computational intelligence to help in designing safe, effective, and sustainable natural remedies to health issues around the globe.

6.9 Conclusion

The use of artificial intelligence in predicting the efficacy and toxicity of phytochemicals is a disruptive development in the modern pharmacological field of study. The use of the AI-driven method has already demonstrated high promise in enhancing the speed of bioactive compound discovery, improving drug-likeness, and increasing drug safety through the combination of molecular descriptors, pharmacokinetics, and biological activity information into a single predictive model. Nonetheless, in spite of these successes, AI usage in phytochemical drug discovery is still significantly limited due to the absence of high-quality and standardized datasets, little model interpretability, and a requirement to be robustly validated in a variety of biological systems. More explainable AI models should be further developed, and they should also incorporate federated and learning strategies to ensure privacy and harmonized data-sharing platforms that enhance the transparency of AI-based predictions, their generalizability, and trust. There is also the need to strengthen interdisciplinary cooperation between chemists, biologists, pharmacologists, data scientists, and regulatory agencies with the aim to guarantee the methodological rigor and regulatory adherence. In this respect, AI can positively influence the gap between the classical medicinal knowledge with the current computational innovation by spearheading the transition of the trial-and-error experimental approach to predictive and knowledge-driven discovery of safer and more effective and personalized phytochemical therapeutics.

References

1. Atanasov, A. G., et al. (2021). Natural products in drug discovery: Advances and opportunities. *Nature Reviews Drug Discovery, 20*(3), 200–216.
2. Arrué, L., et al. (2022). New drug design avenues targeting Alzheimer's disease by pharmacoinformatics-aided tools. *Pharmaceutics, 14*(9), 1914.
3. Bjerrum, E. J. (2017). SMILES enumeration as data augmentation for neural network modeling of molecules. *arXiv preprint*. https://arxiv.org/abs/1703.07076.
4. Babalola, O. O., et al. (2025). Integrating phytochemicals and in silico methods for modern drug discovery: A comprehensive review. *Discover Chemistry, 2*(1), 297.
5. Banerjee, P., et al. (2018). ProTox-II: A webserver for the prediction of toxicity of chemicals. *Nucleic Acids Research, 46*(W1), W257–W263.
6. Chen, X., et al. (2020). Discovery of dual FGFR4 and EGFR inhibitors by machine learning and biological evaluation. *Journal of Chemical Information and Modeling, 60*(10), 4640–4652.
7. Dara, R., Hazrati Fard, S. M., & Kaur, J. (2022). Recommendations for ethical and responsible use of artificial intelligence in digital agriculture. *Frontiers in Artificial Intelligence, 5*, 884192.
8. Fan, J., et al. (2018). Evaluation of SVM, ELM and four tree-based ensemble models for predicting daily reference evapotranspiration using limited meteorological data in different climates of China. *Agricultural and Forest Meteorology, 263*, 225–241.
9. Gilmer, J., et al. (2017). Neural message passing for quantum chemistry. In *International conference on machine learning*. PMLR.
10. Hassan, M. M., et al. (2025). Progress of machine learning-based biosensors for the monitoring of food safety: A review. *Biosensors and Bioelectronics, 267*, 116782.
11. Harvey, A. L., Edrada-Ebel, R., & Quinn, R. J. (2015). The re-emergence of natural products for drug discovery in the genomics era. *Nature Reviews Drug Discovery, 14*(2), 111–129.
12. Jain, A., et al. (2022). Bioenergy and bio-products from bio-waste and its associated modern circular economy: Current research trends, challenges, and future outlooks. *Fuel, 307*, 121859.
13. Jumper, J., et al. (2021). Highly accurate protein structure prediction with AlphaFold. *Nature, 596*(7873), 583–589.
14. Kavlock, R. J., et al. (2008). Computational toxicology—A state of the science mini review. *Toxicological Sciences, 103*(1), 14–27.
15. Liao, Q. V., et al. (2022). Connecting algorithmic research and usage contexts: A perspective of contextualized evaluation for explainable AI. In *Proceedings of the AAAI conference on human computation and crowdsourcing*.
16. Lundberg, S. M., & Lee, S.-I. (2017). A unified approach to interpreting model predictions. *Advances in Neural Information Processing Systems, 30*, 4765–4774.
17. Lu, E.-H., et al. (2024). Evaluating scientific confidence in the concordance of in vitro and in vivo protective points of departure. *Regulatory Toxicology and Pharmacology, 148*, 105596.
18. Liu, M., et al. (2020). Prediction of hERG potassium channel blockage using ensemble learning methods and molecular fingerprints. *Toxicology Letters, 332*, 88–96.
19. Li, M. Y., et al. (2025). Application and research progress on artificial intelligence in the quality of Traditional Chinese Medicine. *Frontiers in Pharmacology, 16*, 1687681.
20. Lee, H., et al. (2025). Recent advances in AI-based toxicity prediction for drug discovery. *Frontiers in Chemistry, 13*, 1632046.
21. Mohanraj, K., et al. (2018). IMPPAT: A curated database of I ndian M edicinal P lants, P hytochemistry A nd T herapeutics. *Scientific Reports, 8*(1), 4329.
22. Newman, D. J., & Cragg, G. M. (2020). Natural products as sources of new drugs over the nearly four decades from 01/1981 to 09/2019. *Journal of Natural Products, 83*(3), 770–803.
23. Qadri, M., et al. (2025). Atezolizumab plus bevacizumab in combination with platinum based chemotherapy for ovarian cancer: A systematic review and meta-analysis of randomized control trials. *Journal of Clinical Oncology, 43*(16_suppl), e17558.
24. Rahman, A., et al. (2024). A comprehensive bibliometric and content analysis of artificial intelligence in language learning: Tracing between the years 2017 and 2023. *Artificial Intelligence Review, 57*(4), 107.

25. Ramsundar, B., et al. (2019). *Deep learning for the life sciences: applying deep learning to genomics, microscopy, drug discovery, and more*. O'Reilly Media.
26. Santa Maria, J. P., Jr., Wang, Y., & Camargo, L. M. (2023). Perspective on the challenges and opportunities of accelerating drug discovery with artificial intelligence. *Frontiers in Bioinformatics, 3*, 1121591.
27. Sarkar, C., et al. (2023). Artificial intelligence and machine learning technology driven modern drug discovery and development. *International Journal of Molecular Sciences, 24*(3), 2026.
28. Saha, U., et al. (2024). Screening of Klebsiella pneumoniae isolates reveals the spread of strong biofilm formers and class 1 integrons. *Journal of Applied Microbiology, 135*(11), lxae275.
29. Singh, S., et al. (2023). Artificial intelligence and machine learning in pharmacological research: Bridging the gap between data and drug discovery. *Cureus, 15*(8), e44359.
30. Souza, E. P. D., et al. (2020). Aplicações do Deep Learning para diagnóstico de doenças e identificação de insetos vetores. Saúde em. *Debate, 43*, 147–154.
31. Subramaniam, K., et al. (2023). A comprehensive review of analyzing the chest X-ray images to detect COVID-19 infections using deep learning techniques. *Soft Computing, 27*, 14219–14240.
32. Vamathevan, J., et al. (2019). Applications of machine learning in drug discovery and development. *Nature Reviews Drug Discovery, 18*(6), 463–477.
33. Wang, Q., et al. (2022). Antibody evasion by SARS-CoV-2 Omicron subvariants BA. 2.12. 1, BA. 4 and BA. 5. *Nature, 608*(7923), 603–608.
34. Wilkinson, M. D., et al. (2016). The FAIR Guiding Principles for scientific data management and stewardship. *Scientific Data, 3*(1), 1–9.
35. Wang, Y., et al. (2025). Modeling protein–ligand interactions for drug discovery in the era of deep learning. *Chemical Society Reviews, 54*, 11141.
36. Yang, X., et al. (2019). Concepts of artificial intelligence for computer-assisted drug discovery. *Chemical Reviews, 119*(18), 10520–10594.
37. Zeng, X., et al. (2020). Target identification among known drugs by deep learning from heterogeneous networks. *Chemical Science, 11*(7), 1775–1797.
38. Zhang, D., et al. (2024). Mm-llms: Recent advances in multimodal large language models. *arXiv preprint*. https://arxiv.org/abs/2401.13601.
39. Zeng, G., et al. (2022). Immunogenicity and safety of a third dose of CoronaVac, and immune persistence of a two-dose schedule, in healthy adults: Interim results from two single-centre, double-blind, randomised, placebo-controlled phase 2 clinical trials. *The Lancet Infectious Diseases, 22*(4), 483–495.
40. Zhu, Z., et al. (2022). Torchdrug: A powerful and flexible machine learning platform for drug discovery. *arXiv preprint*. https://arxiv.org/abs/2202.08320
41. Zhou, C., McCarthy, S. A., & Durbin, R. (2023). YaHS: Yet another Hi-C scaffolding tool. *Bioinformatics, 39*(1), btac808.

Case Studies: AI in Preclinical and Clinical Phytopharmacology

7

Nurhasni Hasan and Isha Shakoor

Abstract

Advances in computing power, machine learning, and big data analytics have made AI a transformative tool for healthcare, biotechnology, and natural product research. In health care, AI enhances clinical decision-making, trial design, patient stratification, and pharmaco-vigilance, enabling precision medicine, integrating multi-omics with patient-specific data. Clinical pharmacologists will be important in guiding the development, evaluation, and deployment of AI tools and ensuring regulatory requirements within frameworks such as the European Commission AI Act are complied with. AI and ML have been comfortably integrated into natural product research, enabling the rapid screening, prioritization, and discovery of bioactive compounds from plants, microbes, and marine sources. Predictive identification of therapeutic molecules, elucidation of metabolic pathways, and diagnosis of plant diseases have been enabled by techniques like deep learning, computer vision, and QSAR modeling. AI has been used to perform metabolite profiling of *Withania somnifera*, evaluate antioxidant flavonoids, and optimize flavonoids extraction from plants in order to enhance yield, reproducibility, and efficiency. AI-driven approaches also find their applications in CADD, high-throughput screening, and development of personalized phytomedicine. Various case studies outline how AI accelerates the processes of antibiotic discovery, such as Halicin; optimizes the formulation of Traditional Chinese Medicine; enhances trial efficiency; and finally offers real-time monitoring of the efficacy of herbal drugs. AI in nanomedicine optimizes nanoparti-

N. Hasan (✉)
Faculty of Pharmacy, Hasanuddin University, Makassar, Republic of Indonesia
e-mail: nurhasni.hasan@unhas.ac.id

I. Shakoor
Department of Plant Protection, Wroclaw University of Environmental and Life Sciences, Wrocław, Poland

Y. Bibi et al. (eds.), *Artificial Intelligence in Phytochemistry*,
https://doi.org/10.1007/978-981-92-0875-3_7

cle-based drug delivery processes and enhances targeting specificity, bioavailability, and controlled release profiles. Challenges for AI in natural product research relate not only to data quality but also to model explainability, integration across disciplines, and compliance with regulatory frameworks. AI holds great potential for further streamlining and facilitating more efficient drug discovery to enhance therapeutic efficacy and solve global health challenges through the identification of new compounds quickly, affordably, and with unprecedented precision. Integration of AI into natural product R&D belongs within the paradigm shift toward data-driven, personalized, and efficient drug discovery. The current chapter presents a description of how AI enables personalized, efficient, and scalable phytomedicine development from bench to bedside.

Keywords

Deep learning (DL) · Neural networks · Phytochemicals · Virtual screening · Molecular docking

7.1 Introduction

Artificial intelligence means performing tasks generally thought to be limited to human intelligence by computational systems. In the last few years, its application in health has grown markedly owing to an increase in computational capacity, increasing availability of complex datasets, and massive progress in ML methodologies. That has resulted in tremendous growth in research and applications based on AI.

Nowadays, AI and machine learning play a transformative role in modern phytochemistry and natural product lead identification. These technological innovations integrate a wide array of datasets-ranging from ethnobotanical knowledge to phytochemical profiling and bioactivity screening—to enhance the accuracy of prioritizing plant species with better data support. Complex patterns and associations might not be obvious through conventional analytical methodologies but could be detected through machine learning algorithms. This brings an opportunity for early identification of bioactive compounds, thereby accelerating the process of discovering novel therapeutic leads [1].

Artificial intelligence is of growing importance in biotechnology, especially within plant molecular pharming, where it serves to enhance yields, stability, and overall product quality. Research could combine AI with synthetic biology and plant-based engineering to improve protein folding and stability, catalytic activity, and organelle targeting [2]. Such algorithms as neural networks, support vector machines, and ensemble regressors analyze training and experimental datasets to design and optimize protein structures for better thermostability, activity, and antibody affinity.

AI is also revolutionizing natural product studies by accelerating the discovery and development of bioactive compounds. Its capability for handling huge datasets, identification of potential molecules, and interpretation of high-throughput screening results make the process of discovery much quicker and more accurate than manual methods. Despite existing challenges, AI continues to significantly advance natural product identification and drug development [3].

Artificial intelligence is now developing computer-aided drug design to be more efficient in developing novel bioactive compounds and diminishing the cost and time it takes to develop new drugs. Using big datasets on natural products, AI will be able to identify important patterns relating chemical structure to therapeutic action, hence giving strong insights into the identification of promising molecules targeting diseases [4]. Thanks to fast improvements in machine learning, researchers are able to widen the bottleneck in natural product discovery and accelerate their search for new therapeutics.

The communicable and non-communicable diseases continue to raise a high health burden globally, despite great efforts invested in developing treatments against conditions such as HIV/AIDS, malaria, hypertension, diabetes, and cancers. Classic pharmaceutical R&D, starting from molecule screening through synthesis or natural sources, followed by its preclinical evaluation, has been successful but unfortunately often missed the goal for these distressing continuing challenges [2]. Revisiting the natural sources repeatedly proved efficient, where drugs such as Taxol, vinblastine, quinine, and artemisinin are of plant origin and remain indispensable in respective therapies. As a result, research around natural products continues to have enormous promise for innovative treatments [4]. In recent years, AI started to be integrated into the field of natural product research and has the potential to completely transform this area. Machine learning algorithms can work with large chemical libraries, find hidden relationships, and even predict biological activities [5]. The AI-driven virtual screening of compounds enables rapid identification of those which can exhibit a desired therapeutic action, decreasing the workload and increasing the efficiency of drug discovery [6]. Moreover, AI can also predict the toxicities, solubility, and pharmacokinetics for major characteristics of compounds to be selected (Table 7.1) [7].

7.2 Data Analysis in AI for Natural Products Discovery and Development

Artificial intelligence helps us in studying huge chemical data sets. Using AI algorithms, it is possible to realize patterns and relationships that are difficult or even impossible to trace by human beings very often [9]. One of the key uses of AI in this regard is the study of chemical compounds' structure to determine the presence of new natural products. For instance, Nath et al. [10] used machine learning to screen large databases of known natural products for new structures displaying potential for biological activity. Such methods of automatic study have enabled AI to process large data sets much more quickly and with increased accuracy than previous traditional methods, while simultaneously recognizing previously unknown chemical structures that have therapeutic potential (Fig. 7.1) [4]. AI also plays a crucial role

Table 7.1 AI applications in natural product research

Domain	AI method	Natural product	Outcomes
Metabolite profiling	Deep learning	Plant extracts	Accurate metabolite classification
Trial design	Bayesian models	Herbal compound	Optimized patient stratification
Real-time monitoring	Reinforcement	Marine metabolite	Adaptive monitoring of biomarkers
Toxicity prediction	Random forest	Microbial product	Reduced false positives
Drug discovery	Neural networks	Alkaloid compound	Faster candidate identification
Quality control	SVM	Fermentation product	Automated purity assessment
Personalized therapy	Decision trees	Botanical extract	Tailored treatment recommendations
Supply chain tracking	Natural Language Processing (NLP)	Essential oils	Improved traceability
Clinical outcomes	Ensemble models	Natural peptide	Enhanced survival prediction

Fig. 7.1 AI application across the phytodrug pipeline

in HTS, wherein hundreds to thousands of compounds are screened for biological activities. Handling enormous volumes of data that emerge from HTS becomes tough, but it can be readily undertaken by AI systems [11]. For example, Pfizer once used AI to devise a Bayesian model of cytotoxicity prediction by evaluating data on approximately 100,000 chemicals [4]. AI-assisted databases form the basis upon which computational modeling is conducted, thereby accelerating natural product discovery and development even further [4]. In computer-aided drug design (CADD), AI has totally changed how new drugs are being found. Traditionally, drug

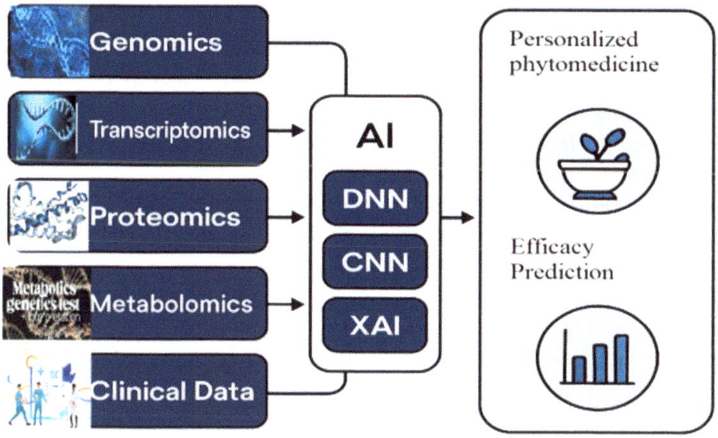

Fig. 7.2 AI integration with multi-omics and clinical data

discovery involved empirical observations followed by random screening. It was a very slow process and lacked efficiency [12]. Subsequently, HTS increased this speed but remained an expensive and time-consuming process; new drug development usually takes about 10–17 years, and it costs about $1–2 billion [13]. With the help of AI-enhanced CADD, particularly Structure-Based Virtual Screening, researchers are now able to screen biologically active compounds with much greater ease, saving a great deal of time and resources in the process (Fig. 7.2) [14]. More recently, novel deep learning architectures have enhanced even further the capacity of AI to predict biological activity from chemical structures [15]. By perusing extensive data about natural products, AI can determine complex relations between chemical structure and therapeutic effect [4]. A potential prediction hence aims researchers at the most promising compounds. This adds to the speed and accuracy with which natural products are found [4]. In general, AI has the potential to innovate the research and study of natural products, facilitating the process of identifying new therapeutic compounds as well as streamlining the process of drug discovery.

7.3 Case Studies

7.3.1 *Withania somnifera* Metabolite Profiling Using AI-Powered Technology for the Preparation of Anti-inflammatory Medication

In clinical plants such as *Withania somnifera* (Ashwagandha), rapid and precise profiling of their compounds is made possible through artificial intelligence (AI), and such technology is currently transforming phytochemical research. The longstanding and numerous therapeutic properties of *Withania somnifera*, a herb native to Southeast Asia and other regions, are due to its various pharmacologically active compounds such as alkaloids, flavonoids, sitoindosides, phenolics, and steroidal lactones. Complex biochemical networks, such as genotype-phenotype combinations, buried within massive datasets generated from contemporary technological progress in high-throughput and omics-integrated research in plant metabolomics, may be determined through deep learning models of computer technology such as Convolutional Neural Networks and Recurrent Neural Networks. Such AI technology improves efficiency in finding key compounds of pharmacological and therapeutic interest concerning drug design and development [1].

Artificial intelligence (AI) is transforming phytochemical research by making it possible to quickly and correctly conduct metabolite analysis in medicinal plants such as *Withania somnifera* (Ashwagandha). Alkaloids, flavonoids, sitoindosides, phenolics, and steroidal lactones are just a few of the known bioactive agents that make *W. somnifera* famous for its medicinal potentials. High amounts of data generated by recent advancements in High-Throughput Metabolomics and OMICS Integration can uncover complex metabolic paths and genotype and phenotype relationships by deep learning techniques such as Convolutional Neural Networks and Recurrent Neural Networks. In this regard, chemicals with probable anti-inflammatory activities were ranked with deep learning for *W. somnifera* metabolomics data. Several candidates were found effectively by these models—most especially withaferin-A—which has been found in cell and animal studies to decrease pro-inflammatory cytokines and increase anti-inflammatory reactions. In this regard, this AI-assisted approach enhanced *W. somnifera* metabolomics understanding and reduced drug discovery and development timelines. One of the best ways to enhance plant-based drug discovery and make it faster and more efficient would be for these technologies to be combined with phytochemical and genomic research [2].

7.3.2 QSAR Modeling for Antioxidant Flavonoids

Artificial intelligence models of QSAR (Quantitative Structure-Activity Relationship) studies conducted using *Citrus sinensis* (orange skin) flavonoids demonstrated good predictability regarding their antioxidant activity. In those studies, high values of correlation coefficients (R^2 = 0.81–0.99) and cross-validated coefficients (Q^2 = 0.82–0.90) were achieved. The modeling process utilized essential molecular descriptors, such as hydroxyl positions, topology, physical-chemical properties, and Fukui indices from density function theory calculations (DFT/B3LYP method), focusing on electronic and structural characteristics essential for radical scavenging ability. Machine learning models, like multilinear regression (MLR), support vector machine (SVM), and gradient boosting machine, were utilized for finding correlations between predicted molecular descriptors and experimental data of antioxidant activity ($\log IC_{50}$ values of compounds tested). The number of datasets ranged from 20 to 46 compounds of flavonoids, which were extensive and diverse in nature to establishing essential SRARs. These results underscored that hydroxyl group modifications in particular positions of a ring and conjugation patterns were extremely important in improving their ability to act as antioxidants, which provided useful inputs for designing effective antioxidants using computational models [3].

7.3.3 AI in Clinical Trial Design for Herbal Formulations

Clinical studies for polyherbal formulations aimed at metabolic syndrome were optimized at Genentech using AI systems. AI classified participants and projected responders vs. non-responders by evaluating patient genetics and lifestyle data, increasing trial efficiency and cutting expenses.

In November 2023, Genentech, a Roche Group company, and NVIDIA signed a multi-year strategic research partnership to use artificial intelligence (AI) to speed up medication discovery and development. Through this collaboration, NVIDIA's cutting-edge accelerated computing and generative AI technologies are combined with Genentech's vast biological and molecular databases, AI capabilities, and research knowledge. The objective is to create a next-generation AI platform that revolutionizes Genentech's existing AI models and algorithms, greatly increasing the precision and speed of finding and creating new treatment options. By accelerating the discovery pipeline and fostering scientific innovation in biomedicine, generative AI has the potential to transform the life sciences and healthcare industry, according to Jensen Huang, CEO of NVIDIA [3].

The partnership strengthens Genentech's continued investment in AI and ML across a range of treatment modalities and research areas. In order to overcome the conventional difficulties of lengthy timelines and high uncertainty in pharmaceutical R&D, Genentech plans to integrate NVIDIA's computational infrastructure to make drug development more predictable, economical, and efficient. In order to improve patient outcomes through the quicker delivery of innovative medications, the company sees AI as a key facilitator for obtaining deeper insights into human biology, disease causes, and target identification. This collaboration emphasizes the intersection of big data, computational science, and molecular biology to revolutionize healthcare innovation, highlighting a larger industry move toward AI-driven precision drug design.

7.3.4 Real-Time Monitoring of Herbal Drug Efficacy

Wearable technology and smartphone apps were integrated with AI to track the responses of patients who took cognitive supplements based on *Bacopa monnieri*. The system tracked sleep, mood, and memory variables to feed data into predictive models that modified dose recommendations in real time.

It is a smart health monitoring system that tracks people's reactions to a herbal supplement produced from *Bacopa monnieri*, a plant believed to enhance memory and focus, using wearable technology (such as smartwatches) and smartphone apps. Important indicators that the system measures include sleep patterns (the quality and duration of a person's sleep), emotional condition, or mood, memory performance (the degree to which an individual can recall information). Artificial intelligence (AI) analyzes all of this data. The AI learns how each person responds to the supplement by searching for trends. Based on this, it may forecast and modify the herbal supplement's dosage in real time; that is, it suggests raising, lowering, or keeping the dosage in accordance with the individual's body and mental state [4]. Major global health issues including diabetes, hypertension, and cardiovascular illnesses put a burden on healthcare systems since they require ongoing, individualized care. Real-time vital sign tracking and predictive analysis using machine learning algorithms are made possible by AI-based health monitoring systems that use smartwatches and mobile applications [5]. AI-assisted monitoring improved patient outcomes, such as lower blood pressure and HbA1c levels and better medication adherence, in a randomized controlled trial that used quantitative and qualitative methodologies [6]. Convolutional neural network (CNN) models, in particular, showed the highest predictive efficiency and accuracy in detecting changes in health and improving illness treatment.

7.3.5 Agentic AI in Pharma: Personalized Phytomedicine

Personalized phytomedicine using Agentic AI in Pharma According to a 2025 White Paper on Agentic AI, pharmaceutical companies make use of autonomous AI agents in controlling the phytodrug pipeline through compound selection down to post-market surveillance. Ginkgo biloba extracts, for example, were tailored using AI based on metabolic markers and cognitive characteristics for senior patients.

AI is revolutionizing personalized medicine through its ability to tailor treatments to individual genetic profiles, hence significantly improving the therapeutic outcome and furthering patient-centered healthcare. Advanced genomic data analysis by AI algorithms identifies genetic markers that predict disease susceptibility and drug response, hence allowing precise and personalized treatment planning. Moreover, AI integrates multi-omics data, including genetic, clinical, and environmental factors, to provide a comprehensive overview of patient health. This not only enhances the clinician's ability to predict which patients will benefit most from a certain therapy but also to develop strategies that maximize efficiency with minimum side effects, reducing trial-and-error in [5]. The presence of AI-driven personalized medicine is witnessed in many medical fields. In oncology, for instance, Tempus uses AI to analyze the genetics of tumors and recommend cancer therapies that are tailored, thus considerably improving survival rates. In neurology, Biogen applies AI to interpret genetic and imaging data for optimizing treatments of neurological diseases such as Alzheimer's and multiple sclerosis. Heart Flow uses AI-based imaging in cardiology to develop unique cardiac models that help in obtaining an accurate diagnosis for intervention. Collectively, these innovations show how AI enhances treatment precision, reduces adverse reactions, and makes healthcare more cost-efficient with evidence-based personalization.

7.3.6 AI for Antibiotic Discovery from Natural Products (Halicin)

In 2020, scientists from MIT and the Abdul Latif Jameel Clinic for Machine Learning in Health designed a deep learning model that could discover new antibiotics from large chemical libraries. The model was trained on about 2500 molecules, comprising FDA-approved drugs and natural products, to learn structural features related to antibacterial activity. It later screened over 100 million compounds for potential antibiotic properties. Among the candidates, it picked out a molecule called Halicin, which originally was investigated as a diabetes drug. This molecule showed previously unrecognized and unique mechanisms of action [7]. Research has showed that Halicin had potent, broad-spectrum antibacterial activity against many multidrug-resistant pathogens, such as *Acinetobacter baumannii*, *Clostridioides difficile*, and *Mycobacterium tuberculosis*. In mouse infection models, Halicin eradicated *A. baumannii* infections within 24 hours, highlighting its strong therapeutic potential. Although Halicin itself was not derived from any single natural product, this AI-driven discovery illustrates how one can apply deep learning in natural product research to surmount the "dereplication problem"—a frequent rediscovery of already known compounds. By extending similar AI models to natural product databases, researchers can quickly identify structurally novel and bioactive compounds, thereby accelerating antibiotic discovery and heightening the search for new therapeutics from natural sources.

7.3.7 AI-Accelerated Cancer Drug Discovery from Traditional Chinese Medicine (TCM)

Traditional Chinese Medicine supplies a wide and diverse reservoir of potential drug candidates, considering that it has long utilized natural products for therapeutic purposes. However, the complexity of the formula in TCM and the limited systematic investigation have been presenting some crucial challenges for efficient drug discovery and development. Of late, AI has become a transformative tool that is helping to accelerate drug discovery based on TCM by integrating and analyzing large-scale data involving herbal properties, chemical compounds, and clinical applications. Machine learning and deep learning algorithms have been particularly effective in bioactive compound identification, prediction of drug-target interaction, and elucidation of the mechanisms of action of complex herbal combinations [8]. Network pharmacology, which, when coupled with AI, enables the mapping of multi-target and multi-pathway interactions intrinsic to TCM formulations, allows for the elicitation of therapeutic effects at the systems level. Artificial intelligence-based compound screening and drug repurposing does not lag behind either. This further assists in unearthing new usages of medicinal herbs [9]. However, there remain some issues to be addressed regarding supplying data standardization, interpretability, and incorporation of knowledge from conventional practices in a modern pharmacological paradigm. Improvement in this direction can enable the maximum utilization of artificial intelligence for enhancing the drug discovery of Traditional Chinese Medicines on this front as well [10].

7.3.8 AI for Optimizing Extraction of Flavonoids (Quercetin and Kaempferol)

AI models improve the extraction of these useful compounds from plants by analyzing specific parameters for type of solvent, temperature, and pressure to attain the maximum yield. A review in Foods in 2024 noted the trend that keeps increasing: applying algorithmic and machine learning approaches for the optimization of flavonoid extraction. In fact, they discuss how such procedures increase very remarkably in efficiency and reduce material waste [11].

Another research group [12], on the other hand, worked on the optimization of flavonoids extraction from Polygonatum using a backpropagation ANN, where the extracted yield was found to be higher when compared to traditional methods. Such AI-based approaches ensure that the preparation of active ingredients is of reproducible quality, hence forming a key basis for drug discovery and development.

However, while promising at a preclinical and laboratory scale, there is a critical lack of clinical or post-market case studies that have demonstrated a full pipeline from AI-optimized extraction right through to clinical validation or commercialization. Reviews, such as those in *Separations*, 2024, and *International Journal of Molecular Sciences*, 2024, discuss the pharmacological potential of flavonoids and the advantages of advanced extraction technologies but note that integrating AI into large-scale regulated pharmaceutical production is still emerging. Thus, even though AI-driven extraction optimization is a powerful step forward in ensuring quality and efficiency in the early parts of the drug discovery supply chain, further research is needed along with clinical validation and regulatory alignment to see these advances translated into real-world therapeutic and market applications.

7.3.9 AI-Enhanced Clinical Trial Design and Patient Selection

AI changes the manner in which clinical trials take place, making them faster, safer, more efficient, and better designed, executed, and monitored. AI allows the selection of patients, forecasting of outcomes, and monitoring in real-time through machine learning and predictive analytics, hence more rapid decision-making. These technologies mine huge amounts of data for trends that allow researchers to construct far better trials with fewer risks in less time 12. For example, AI can rapidly go through hundreds of health records of patients, shortlisting potential test participants to speed up the recruitment process while allowing AI to optimize trial designs to accommodate fewer test patients. Other AI Capabilities during the trial process include constant health data monitoring of patients in real-time, hence the prompt detection of adverse reactions, boosting safety, and perfect adherence. AI further handles the complex data outcome of clinical trials much faster and with high precision compared to human methods, hence high efficiency. Practical examples of AI applicability have impacts on how clinical trials are conducted. Pfizer uses AI in predictive modeling of possibilities for patients' accrual and dropout rates, further optimizing timelines and maintaining tighter control of the purse strings 12. AI technologies make use of natural language processing, thus rapidly shortlisting potential test candidates out of voluminous health records that are suitable for all kinds of trials. Antidote Technologies uses AI for trial patients' allocation across the globe, further boosting test participation rates and trial success rates. AI technologies have, however, not lost significant ground on data secrecy, trial issues, and the demand for proper regulation. AI use faces ethical concerns regarding biases and impacts on equity of the trial process using AI. Anyhow, AI implementation within clinical trials will result in lowered expenses, expedited development, increased safety, so AI will make the revolutionary approach toward new treatment developments, as well as personalized treatments for patients.

7.3.10 Predicting Herb/Drug Interactions Using Artificial Intelligence

The possibility offered by herb–drug interactions (HDIs) has substantially increased due to the constantly increasing use of herbal supplements in traditional health care facilities. Indeed, the traditional pharmacovigilance system is generally dependent upon inconclusive post-marketed observations. With the intention to meet these hurdles, scholars at the University of California, San Diego, and the Scripps Research Institute developed an AI system to predict HDIs using natural language processing based upon medical literature and electronic health records. With the intention to spot possible interactions before the occurrence of actual HDIs, these models examine patient data, pharmacokinetics, and molecular shapes. For instance, models based upon artificial intelligence established interactions among selective serotonin reuptake inhibitors (SSRIs) and St. John's wort to encourage health practitioners to avoid detrimental occurrences. Such strategies encourage tailor-made medicine projects.

7.4 Use of Artificial Intelligence in Quality Control and Authentication of Herbal Products

With the increasing demand for herbal medicines and use of various materials in processing these products, there has always been a challenge in terms of authenticity and guaranteed quality for these herbal medicines. To attain authenticity in these herbal products such as Ginseng and Curcuma Longa, for instance, machine learning techniques were used by research scientists in the National Institute of Food and Drug Safety Evaluation in South Korea. Developing highly precise models for geographical identification, phytochemical consistency, and authenticity of these herbal medicines using Convolution Neural Networks and Support Vector Machines, artificial intelligence-powered mechanisms for quality analysis are much superior in terms of speeds and accuracy over existing processes, which has been discussed in a report released by Korea Institute of Science & Technology in 2024.

7.4.1 Mining of Ethnopharmacological Knowledge Using Artificial Intelligence

Despite the abundance of ethnopharmacological traditional knowledge available and suited for modern drug developments, still much of traditional knowledge in ethnopharmacology has not been organized and harnessed. The use of AI technology in processing information from ethnobotanical literature and traditional handing down has been led by the Swiss Federal Institute of Technology, known as ETH Zurich, in partnership with the University of Ghana. From ethnobotanical knowledge, scientists have been able to discover new medicinal plant species that have the capability of developing new medicines via machine learning approaches of digitized manuscripts. Examples of such developments involve the use of AI technology in guiding traditional knowledge of medicinal plant selection for malaria drugs in West Africa, based on traditional rates of phytochemical analysis.

7.5 Conclusion

Artificial intelligence has already proved to have potential in revolutionary applications in various preclinical and clinical aspects, including pharmaceutical discovery and development, target identification, predictive toxicology, and diagnosis, prognoses, and treatment strategies. New evolving models of federated learning, use of real-world evidence, and future validation studies are expected to further boost the reliability and relevance of AI models in real-world applications. Nonetheless, effective translation of AI models for use in real healthcare settings demands effective teaming between biologists, physicians, technical experts, and policymakers. Consistent regulatory frameworks are also essential to ensure privacy, transparency, and patient safety.

References

1. Nag, S., et al. (2022). Deep learning tools for advancing drug discovery and development. *3 Biotech, 12*(5), 110.
2. Abdelwahed, M. T., Hegazy, M. A., & Mohamed, E. H. (2023). Major biochemical constituents of Withania somnifera (ashwagandha) extract: A review of chemical analysis. *Reviews in Analytical Chemistry, 42*(1), 20220055.
3. Wang, X. (2025). *Industry innovation in the era of artificial intelligence: The AI compass.* CRC Press.
4. Sharma, A., et al. (2024). Herbal nutraceutical as alternative medicine. In *Herbal nutraceuticals: Products and processes* (pp. 23–45). Wiley.
5. Gupta, P., & Pandey, M. K. (2024). Role of AI for smart health diagnosis and treatment. In *Smart medical imaging for diagnosis and treatment planning* (pp. 23–45). Chapman and Hall/CRC.
6. Ma, S., Zhang, M., Sun, W., et al. (2025). Artificial intelligence and medical-engineering integration in diabetes management: Advances, opportunities, and challenges. *Healthcare and Rehabilitation 1*, 100006.

7. Wang, S., et al. (2024). Halicin: A new horizon in antibacterial therapy against veterinary pathogens. *Antibiotics, 13*(6), 492.
8. Li, C., et al. (2022). Multi-compound and drug-combination pharmacokinetic research on Chinese herbal medicines. *Acta Pharmacologica Sinica, 43*(12), 3080–3095.
9. Bohr, A., & Memarzadeh, K. (2020). The rise of artificial intelligence in healthcare applications. In *Artificial Intelligence in healthcare* (pp. 25–60). Elsevier.
10. Nath, R., et al. (2024). The importance of natural products in drug discovery: opportunities and success. In *Traditional resources and tools for modern drug discovery: Ethnomedicine and pharmacology* (pp. 351–376). Springer.
11. Dara, S., et al. (2022). Machine learning in drug discovery: A review. *Artificial Intelligence Review, 55*(3), 1947–1999.
12. Thomas, M., et al. (2021). Applications of artificial intelligence in drug design: Opportunities and challenges. In *Artificial intelligence in drug design* (pp. 1–59). Humana New York.
13. Imrie, F. (2021). *Deep learning approaches for pre-clinical drug discovery*. University of Oxford.
14. Chopra, H., et al. (2022). Application of artificial intelligence in drug discovery. *Current Pharmaceutical Design, 28*(33), 2690–2703.
15. Selvaraj, C., Chandra, I., & Singh, S. K. (2022). Artificial intelligence and machine learning approaches for drug design: Challenges and opportunities for the pharmaceutical industries. *Molecular Diversity, 26*(3), 1893–1913.

Part IV

Integration of AI in Phytopharmacological Research

AI Techniques in the Discovery of Novel Phytochemical Compounds

8

Rahmat Wali and Marcello Iriti

Abstract

The use of natural compounds for the treatment, prevention and management of various diseases is as old as the origin of humans. The subsequent discoveries of bio-active compounds occurred in the nineteenth and twentieth centuries, which developed the foundation of phytochemistry. Phytochemicals are bioactive compounds with pharmacological properties. They offer health benefits and have high demand across industries, including pharmaceuticals, nutraceuticals, cosmetics, etc. The complexity of phytochemicals hinders their isolation, purification and characterization. In addition, their low stability, lower yield, and toxicity also pose limitations in their comprehensive analysis. Traditional methods like spectrometry and chromatography have been used for the discovery of phytochemicals, but they are time-consuming and have low detectability. AI transformed phytochemistry by enabling efficient structural elucidation and metabolite profiling of compounds. They process large-scale data sets with high accuracy. It speeds drug discovery by screening thousands of compounds in a short duration to predict their structure and biological activities. The algorithms use supervised and unsupervised data and can identify novel compounds. This chapter provides an overview of the origin of phytochemistry and how AI solutions are transforming conventional methods into data-driven, optimized, and expeditious processes.

R. Wali (✉)
Barari International Limited Company, Specialist Plant Taxonomy & Biodiversity, Section-Barari KSA-Branch, Riyadh, Kingdom of Saudi Arabia

M. Iriti
Department of Biomedical, Surgical and Dental Sciences, University of Milan, Milan, Italy

Y. Bibi et al. (eds.), *Artificial Intelligence in Phytochemistry*,
https://doi.org/10.1007/978-981-92-0875-3_8

Keywords

Phytochemicals · Bioactive compounds · Artificial intelligence (AI) · Drug discovery · Metabolite profiling

8.1 Introduction

Phytochemistry has been traditionally relied on rigorous testing of plants and their chemical compounds through bioassay-guided fractionation and expert driven interpretation of analytical data. This process has yielded many drugs, but due to a large amount of digital information about plant phytochemicals, metabolomic datasets, genomes and transcriptomes, and vast literatures in phytochemistry and ethnobotany, this procedure has intrinsically low throughput and is biased toward well-studied species and chemotypes. However, the limitation in scalable data generation is no longer a bottleneck. AI enables us to mine literature on plant compound relationships, recognize spectral patterns without hand crafted rules, propose new molecular scaffolds, and integrate multi-mics for highlighting biosynthetic hotspots. Deep learning and large learning models (LLMs) have thus far illustrated that complicated scientific patterns, like enzyme Product pairs, compound-species associations, and NMR features, can be learned in a direct fashion from raw data with high accuracy [1].

The incorporation of AI methods into phytochemistry is currently transforming the way hypotheses are formulated and evaluated. Large-scale language models, when trained on scientific text corpora, are able to rapidly integrate the associations among ethnobotanical data, metabolomics, and genomics, thereby revealing hidden patterns that would be difficult to discern within the existing paradigm of discovery. Moreover, deep learning models are able to predict the pathways of biosynthesis, thereby providing key perspectives on the enzyme promiscuity, structure, and function within ecological domains [2]. This hybrid framework of computational models and biological discovery, therefore, not only accelerates the pace of compound discovery but also ensures equal access to undescribed taxa, thereby limiting the role of ancestral species within phytochemistry as the paradigm shifts to include faster, more scalable, and reproducible models of discovery.

This chapter provides information about AI techniques used for literature and ethnobotanical mining, operating upstream of target binding and docking, spectral pattern recognition, generative scaffold design, untargeted metabolomics, graph-based discovery, and multi-omics fusion.

8.2 AI-Powered Literature Mining and Ethnobotanical Intelligence

8.2.1 NLP Foundations for Phytochemical Lead Mining

A large quantity of phytochemical knowledge is still unexplored. As they reside in unstructured sources, including plant chemistry reports, ethnobotanical surveys, and pharmacological studies, natural language processing (NLP) allows the automatic extraction of key entities, that is, plants, metabolites, enzymes, biological activities, and the relationships between them. Domain-specific transformer models, which have been pretrained on large scientific literature, for example, Materials BERT [2], have shown that pretraining on millions of abstracts greatly increases the accuracy of entity identification and relationship extraction compared to general models. For instance, one such pipeline extracted ~300,000 material-property records from 130,000 abstracts in just 60 hours, demonstrating how such approaches may be scaled for plant-compound mining [3]. In the study of specialized plant metabolism, text mining has been applied to compile large datasets of compound species associations, which are then mapped onto molecular phylogenies to generate phylochemical maps. In tyrosine-derived natural products, mining and manually curating >2500 articles yielded a high-quality dataset that revealed lineage-specific distributions such as betalains and Amaryllidaceae alkaloids and clades enriched in dopamine-derived compounds [3]. These maps indicate plant groups that are likely to contain novel analogs and unusual biosynthetic pathways, thus informing future discovery efforts.

Beyond just indicating the association of compounds with species, there has also been an increased application of AI-based text mining and deep learning techniques in conjunction with spectral and genomic data with the aim of establishing integrative discovery tools for compounds. In this way, the combination of compound annotations derived from NLP techniques and metabolomic fingerprints has enabled the automated identification of structural categories and biosynthetic pathways, while the association of the metabolomic maps with the transcriptomic maps has enabled the identification of enzymes and gene clusters essential in catalyzing the compounds in metabolome of plants.

8.2.2 Large Language Models and Ontology-Driven Extraction

Large Language Models (LLMs) are an extension of classical NLP technique, as they simultaneously perform the task of compound identification and relation extraction to provide organized records rather than fragmented results. For instance, the (Generative Pre-trained Transformer 3) GPT and LLM models, fine-tuned with small amounts of annotated data, achieve a level of accuracy at 69–95% for intricate chemical text-mining problems. Such problems include the identification of chemical names, the determination of NMR settings, and the transformation of unstructured experiment procedure descriptions to organized action sequences [4]. For materials chemistry, the GPT-3 and LLM2 models have been employed to determine the composition, properties, and applications and simply provide the information, organized as JavaScript Object Notation (JSON) facilitating the autonomous creation of Knowledge Graph (KGs) [5].

LLMs have been used in plant metabolism studies regarding extracting enzyme-product pairs, compound-species relationships, and numerical information from tables [6]. By leveraging retrieval-augmented generation (RAG) together with LLMs and appropriate prompt designs, these models reached a level of 80–90%

accuracy in recognizing enzyme-product pairs and improving false-negative outcomes in extracting compound-species relationships relative to previous models based upon rule-coding [6]. These strategies can be extended to either phytochemistry or ethnobotany in creating plant-compound-use data repositories through far less human curation work.

However, post-processing is still a crucial step. The entities derived from text analysis have to be normalized based on the accepted taxonomies, metabolite names, and medical and/or pharmacological terms. In the context of phylochemical mapping, for instance, pattern mining was used to identify potential associations, which were then authenticated through reference sets before the training process for the LLM [1]. The use of LLMs for processing can mechanize this normalization process, with low false positives but significant false negatives [6]. More recent work has demonstrated the application of multimodal LLMs, where text is integrated with spectral and omics information, for the concurrent learning of chemical structure, biosynthetic modules, and ecological roles [2]. It appears, therefore, through these different advances, the potential for LLMs exists to revamp the science of phytochemistry.

8.2.3 Ethnobotanical Intelligence and Multimodal Extraction

Much ethnobotanical knowledge is present in the form of tables and figures within PDFs, where they are not easily accessible by traditional text-mining approaches. These papers have now become available through transcription of such tables to machine readable data formats facilitated by multimodal LLAMs that can handle images alongside text. A recent case study demonstrated the use of GPT-4–powered multimodal workflows to extract quantitative phytosterol abundances for 67 plant species from 13 complex table images. In their raw state, the tables, each with diverse layouts, variously mixed units, and summary rows, yielded initially very low accuracy. Application of some light human preprocessing, including the removing of misleading elements, harmonization of species names, etc., and a two-stage pipeline of image transcription followed by chemical/species sanity checks remarkably increased performance.

This workflow exemplifies a broader pattern, i.e., modest human curation augmented with multimodal LLM transcription can transform scattered ethnobotanical tables into structured datasets (Fig. 8.1). A similar approach has been adopted in materials science, where multimodal GPT-4 pipelines captured composition property application tuples directly from figures and tables, enabled automated knowledge graph construction. For plant sciences, multimodal LLMs have also been tested on extracting enzyme product pairs and quantitative data from spectral plots, performing at 75–85% accuracy when combined with retrieval-augmented generation.

In ethnobotanical surveys, for instance, it can automatically capture species names, traditional uses, preparation methods, and known chemical constituents from tabular records. Standardization of these by ontology guided post processing

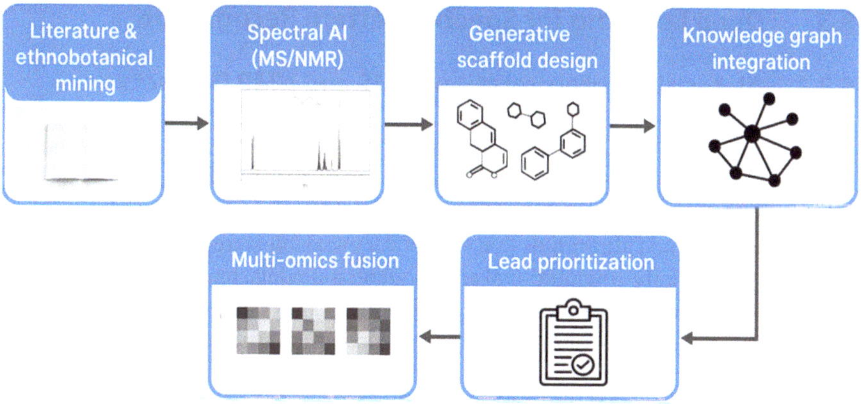

Fig. 8.1 Workflow of AI-driven phytochemical discovery

will render such datasets amenable to chemotaxonomic mapping, comparative phylogenetic analyses, and AI-driven lead prioritization. Such an integrative strategy now holds much promise for accelerating the discovery of novel plant metabolites and underlined hitherto unexplored taxa with unusual biosynthetic potential [1].

8.2.4 Toward Comprehensive Phytochemical and Ethnobotanical Knowledge Graphs

The combination of domain-specific transformers has fine-tuned large language models, multimodal extraction techniques, and ontology-driven normalization that has empowered the creation of large-scale phytochemical and ethnobotanical knowledge graphs. Such approaches have already proven to be fruitful in the realm of polymers and material sciences, where automatic pipelines were employed for the creation of structured databases able to facilitate non-trivial analyses and hypothesis building [5]. Phylochemical maps generated by text mining and large language models are increasingly offering the backbone for the integration of omics, ecological, and ethno-data in the plant chemistry field [6]. Such knowledge graphs can have multiple applications, including clade identification with exceptional levels of chemical diversity to inform focused isolation efforts [1].

The prediction of likely species possess analogs of known scaffolds, according to shared phylochemical and/or ethnomedicinal profiles [3]. Furthermore, recognizing inconsistencies between traditional use and established chemical knowledge may inform hypothesis-driven metabolomics or pharmacological investigations.

8.3 Deep Learning for Spectral Pattern Recognition

8.3.1 AI-Assisted Interpretation of MS and NMR Datasets

Phytochemical instrumentation such as LC-MS, GC-MS, and NMR produces highly correlated and high-dimensional data, wherein real molecular peaks are easily confounded with noise and other peaks. Traditional approaches are dominated by hard-programmed algorithms that are not very scalable or reproducible. Recently, AI models, especially convolutional and sequential models, have proved highly successful as an alternative approach because these models can automatically identify real peaks, predict functional groups, and reconstruct spectra using time-domain information.

Within the context of LC-MS metabolomics, convolutional neural networks (CNNs) have been used to differentiate true and false spectral peaks, successfully filtering out up to 90% of false peaks with zero loss of true positives, hence improving data reproducibility across datasets [7]. A different CNN algorithm for peak detection, this time dedicated and specifically trained on high-resolution data from LC-MS ("PeakOnly"), nearly reached the optimum accuracy by adaptively filtering out noisy peaks of low intensity [8]. More sophisticated CNN and Faster R-CNN models have processed two-dimensionally resolved ion chromatograms as images (Peak_CF), attaining >90% accuracy and recall on test datasets and outperforming current state-of-the-art software in F1-score and robustness to severe chromatographic drift [9]. Confronting NMR spectrocopy data, deep neural networks have been used both for advanced peak-picking and radical approaches to revolutionizing current data processing. NMR spectral data can be accurately picked using deep neural networks, effortlessly extending the limits of human experts and finally resolving the historical inability of computerized approaches [10]. WaveNet models, this time trained on non-uniform data, have accurately reconstructed NMR data interpreted at highest resolution by identifying specific point spread patterns associated with individual peaks, outperforming state-of-the-art approaches [11]. Even more recently, neural nets have recently shown their potential to directly transform free induction decay (FID) signals into spectra and lists of peaks, hence perfectly substituting Fourier transformation and even resulting in virtual infinite resolution spectra within the context of a related proof of concept [12].

Another frontier area could be functional-group predictions. For example, the use of a multi-label neural network based on FTIR and MS fingerprints successfully predicted the identification of all functional groups in pure substances and mixtures, without requiring programmatic rules based on expert knowledge or matching peaks within libraries, and was experimentally verified to work successfully, proving the ability of spectral deep learning to "think like a spectroscopist expert" [13].

These accomplishments illustrate the potential for deep learning to revolutionize spectral analysis as it applies to phytochemistry and metabolomics research. These algorithms, collectively, diminish noise, increase the potential for reproducible results, and offer the possibility for the automated prediction of functional groups, making such analysis comparable to or perhaps beyond the analysis done by experts today and offering new opportunities for the identification of compounds.

8.3.2 Automated Peak Annotation, Clustering, and Mixture Analysis

For peak annotation, artificial intelligence tools can be employed for annotating or clustering the spectra, which may assist with compound dereplication. Artificial intelligence algorithms have also been developed for the localization of peaks, classifying peaks as target or non-target peaks (metabolites of analytical interest) within the scope of LCMS, without the aid of component deconvolution software, based on data-driven learning of peak features inherent to the technique [9]. For targeted screening procedures, such as equine doping, CNNs employed on the image pixels of the chromatogram have displaced the endless tedious manual review with zero false-negative results, for the purposes of equine doping analyses [14]. In the case of NMR spectroscopy, there are instruments like SMART 2.0, which use the capabilities of deep learning and are able to swiftly identify the mixture spectra against a compound library. This helps in the swift identification of known skeletons and the filtering of fractions for novel compounds [15]. When used for a marine cyanobacterial extract, it enabled the rapid discovery of a swinholide-like macrolide and novel peptides [15].

Taken together, these examples represent the ways in which spectral deep learning can be integrated within natural product discovery efforts. By the use of automated annotation and clustering, not only has the dereplication process notably been made easier, but the triaging of complex samples has now enabled scientists to allocate resources to those samples that are most likely to yield unprecedented chemotypes.

8.3.3 Model Architectures: CNNs, Transformers, and End-to-End Learning

The spectral data could be expressed in several forms, including images chromatograms, one-dimensional signals, or pseudo-sequences; therefore, various deep learning architectures could be applied.

CNNs CNNs are notably proficient at local pattern recognition and have seen widespread utilization in spectroscopy for tasks as varied as peak detection, classification, and quantitative calibration [8]. These CNNs, with their end-to-end architectures like DeepSpectra, can learn directly from raw data for visible and near-infrared spectra. Such models mostly outperform the pipeline-based models using extensive preprocessing and classical chemometric methods. Performance further improves as training datasets grow larger.

Transformer-based models Transformers have recently been introduced for spectral analysis tasks that require modeling complex mixtures. For example, a hybrid CNN–Transformer algorithm (RST) transforms spectra into "word vectors" and applies attention mechanisms to assign contribution weights. This approach attained almost perfect identification rates (>99% recall) in Raman mixture analysis and provided improved interpretability compared to traditional CNNs [16].

WaveNet-like and sequence models Such architectures take advantage of the temporal structure inherent in FID signals or Non-Uniform Sampling patterns in NMR by learning an instrument-specific point spread function. This results in superior reconstruction and even virtual homo-decoupling that goes beyond common reconstruction techniques [12]. Applied to the phytochemical discovery of, for example, plants metabolites, these models are being trained with labels containing compound identities, structural classes, or biological activities. Once trained on plant-derived metabolomic and NMR datasets, one of the most salient advantages is their generalization across large batches of spectra, allowing for very efficient annotation of new extracts. This scalability allows decreasing the dependency on manual interpretation and supports dereplication, therefore the rapid identification of novel compounds. Deep learning architectures will thus promise to change phytochemistry, as more raw spectral data becomes available, into an automated, high-throughput discipline connecting raw spectral data with compound identification pipelines.

8.3.4 Model Architectures: CNNs, Transformers, and Sequence Learning

The spectral information can be represented in various forms, such as images (chromatograms), one-dimensional arrays, or pseudosequences, which can accommodate various neural network models for analysis.

Convolutional Neural Networks (CNNs) CNNs have proven to be highly effective for local pattern recognition and have seen success in various spectroscopy-related tasks such as peak identification, classification, and quantitative analysis [8]. End-to-end CNN models such as DeepSpectra have been successful in direct analysis of visible and near-infrared spectra without requiring any primary preprocessing of data, and their performance even increases with expanding training data sets, indicating their scalability with regard to spectral analysis of large data sets [17].

Transformer models Transformers were recently applied for spectral tasks, whose complexities include mixtures. A hybrid model using CNN and a Transformer, named RST, converts spectra to "word vectors" and used attention models for attribution weights. Such a model established near-perfect classification accuracy (>99% recall) for Raman mixture classification, with better explainability than a standard CNN, indicating the effectiveness of the attention models for spectral learning [16].

WaveNet-like and sequence models Such models leverage the FID or non-uniform sampling patterns found within NMR. Through learned instrument-specific point spread functions, the reconstruction and even homo-decoupling tasks can be accomplished with greater success than more conventional reconstruction methods [12].

When applied to phytochemicals, the architecture can be trained on metabolomic and NMR spectral data from plants with labels such as the chemical identity, class, or bioactivity. The models created can then be applied to large numbers of spectral datasets, allowing for efficient annotation and dereplication, as the need for human interpretation will be reduced. As raw spectral data continues to be gathered, architectures like deep learning are expected to revolutionize the field of phytochemistry, allowing for the direct identification of compounds from spectral information.

8.3.5 Integrating Spectral AI into Phytochemical Workflows

With the advent of AI-driven phytochemical discovery, spectral models can undertake several pivotal tasks to seamlessly transition from data to valuable chemical information:

High-fidelity peak detection and de-noising Deep learning models are able to increase the quality of untargeted metabolomics datasets (LC-MS/GCMS) by correctly identifying true chemical features among noise and artifacts. This greatly facilitates the process of clustering and statistical analysis, resulting in credible feature sets [8].

Inference on functional groups or scaffolds Through the integration of data from different spectral modalities, AI approaches are able to predict structural characteristics like functional groups or partial scaffolds. While these predictions might not result in the dereplication of the entire compound, they remain useful from a structural standpoint [15].

Mixture analysis and novelty discovery Using the capability of Spectral AI, unusual or under-represented patterns within certain fractions will be identified, which will prioritize them for purification and structural analysis, thus expediting the discovery of novel metabolites [15].

High-throughput screening Automation—The task of examining chromatograms can now be performed by methods involving CNN, thus leaving the chemists with the remaining samples that are difficult to interpret by the model, thus saving time while retaining accuracy [14].

Although these improvements have been achieved, there are still areas that require improvement in this field. First, the construction of large-scale labeled datasets for plant-derived compounds in chemistry would be crucial in enhancing generative capabilities in this domain. Variability in instruments as well as in a chemistry lab needs to be handled in a way that sustains predictability across all platforms. Another area that would be important in chemistry would be interpretability related to predictions from AI systems in chemistry. However, work in analytical chemistry, metabolomics, and natural products research has confirmed that deep spectral learning can be effectively used to facilitate the processing of spectral data from raw form to meaningful representations that are relevant to chemistry. This technology will ensure that phytochemical research processes are scalable and discovery-oriented.

8.4 Generative AI in Novel Scaffold Design

8.4.1 Molecular Representations for Phytochemical Generation

These models take the structured representation of molecules as input and infer the distribution of known molecules and generate new ones. The most common representations are SMILES, with the improved variant SELFIES and the representation as molecular graphs. NLP-inspired models view the representation of SMILES as "language," where tokenization and sequence models are employed to model the syntax and semantics of the chemical representation [18]. Graph models, instead, are designed to directly model the atoms and bonds, thus easily incorporating the topology.

For phytochemistry, training data could be compiled from natural product databases and plant-specific metabolite catalogs. The need to incorporate coding schemes that retain stereochemistry, complex ring systems, and chiral centers arises, given the structural diversity of phytochemicals.

Although this approach could be adopted for phytochemistry, incorporating active compounds from pharmacology via physiological databases and pharmacological catalogs that could enhance research in related domains such as chemoinformatics and bioinformatics, which require robust data integration and analysis.

8.4.2 VAEs, GANs, and Transformer Generators

Variational autoencoders (VAEs) learn to encode the molecules into a continuous latent space, where new molecules can be generated to be decoded into valid molecules. When VAEs are trained on natural products, the latent space captures the "natural product-like" properties like stereochemical complexity and heterocyclic diversity. These properties allow the exploration around known scaffolds with controlled variations. Generative Adversarial Networks (GANs) and Autoregressive Transformers are able to directly parameterize the distribution over the training structures and generate valid SMILES strings or molecule graphs. In the context of drug discovery, such models have been employed for the generation of novelty chemotypes with defined physicochemical properties and bioactivity fingerprints [18]. For the phytochemical scenario, the model could be conditioned on plant lineage, biosynthetic type (e.g., alkaloids and flavonoids), or partial scaffolds derived from spectral information, offering the flexibility of generation for defined plant families. Transformer-based chemical language models: These models can be fine-tuned for generation tasks with certain limitations once they have been trained on large SMILES datasets. These models are applicable for suggesting analogs of a certain phytochemical compound, finalizing structures that are predicted based on MS/MS information or generating lineage-specific scaffolds [18]. The excellent sequence modeling ability often leads to high validity, diversity, and novelty in the generated compounds, which are very useful for the expansion of phytochemicals.

8.4.3 Reinforcement Learning for Scaffold Ref

Although generative models are able to search chemical space in a rather broad manner, there are various reinforcement learning (or RL) models that are used in guiding molecular generation towards a targeted objective. These involve a generative model that presents candidate molecules to an external "oracle," namely, a QSAR model or a properties estimator, which rewards them based upon their merit related to properties like drug likeness or ADMET profiles.

Concerning phytochemical research, the RL can be adapted to multi-objective approaches, which would consider the distinctive nature of natural products:

- Conserving natural product character and innovating: Maintaining the stereochemical complexity, heterocyclic diversity, and other features that distinguish plant metabolites and innovating into new chemical space
- Improving biosynthetic feasibility: Guiding generation toward molecules with a biosynthetic plausible route from known biosynthetic precursors or pathways
- Balancing drug-relevant constraints: Maximization of pharmacological properties while avoiding the extreme of highly synthetic-like chemical structures that are no longer related to natural product scaffolds

Another aspect in which generative policies might be biased is toward regions of chemical space linked to certain types of plant lineages and/or chemotypes. This allows scaffold hops to be performed either within or between plant lineages, facilitating the identification of analogs and new compounds that are biologically and biosynthetically meaningful. A lineage-sensitive approach to reinforcement learning might thus help to quickly find biologically interesting phytochemicals by leveraging the exploratory and fine-resolution abilities afforded by generative models.

8.4.4 Role and Scope in Phytochemical Pip

"Generative AIs must not be considered substitutes for experimental discovery, but as tools that aid phytochemical research. Within this view, generative models may aid in phytochemical research in the following ways":

- Virtual analogs: Proposing analogs for newly discovered natural product skeletons that can be synthesized by semi-synthetic or total synthesis, thus increasing the chemical diversity derivable from natural sources
- Elucidation of structure support: Suggesting plausible molecular structure consistent with limited data, like accurate mass value or functional group attribution, to assist chemists with choosing structure elucidation strategy
- Investigation of latent phytochemical space: Focusing on scaffold clusters that are presently underrepresented in natural product collections and thereby enabling the discovery of novel phytochemicals and their biosynthetic routes.

Notably, this chapter steers clear of docking predictions and target binding predictions. However, generative models are introduced for exploring and expanding structural diversity for hypothesis generation, exploration, and prioritization purposes rather than optimizing molecules for biological targets. Other chapters will explore those concepts in more depth.

8.5 AI in Untargeted Metabolomics and Compound Prioritization

8.5.1 High-Dimensional Profiling and Dimensionality Reduction

Untargeted metabolomics of plant extracts will typically yield thousands of spectral features across multiple samples, ranging from species to treatments. The challenge of interpreting such a high dimensional space is addressed using analysis techniques that are able to maintain meaningful biological structure in reduced dimensions. Non-linear dimensionality reduction tools such as t-SNE and UMAP are popular for mapping complex LC/MS and/or GC/MS data to two and/or three dimensions for exploratory data analysis, and these data visualizations will yield clusters of samples representing chemotypes, species, and treatment conditions.

Also, deep learning methods can be employed in combination with the above techniques to improve the interpretability of the results even further. Using joint embeddings, derived by amalgamating the spectral data with the underlying metadata, improves the coherence of the clusters and facilitates the visualization of the outliers/metabolomic profiles in the data set.

> The combination of image analysis, using 'chromatographic images', with techniques of manifold learning can extract local spectral information with global constraints, while improving the robustness of metabolomic data visualization and clustering [9].

Such integrated workflows not only aid in exploratory data analysis but further help in related applications like chemotaxonomic mapping, comparative analyses in the ecological domain, along with the prioritization of novel metabolite patterns for isolation. As the size of the datasets increases, a combination of dimensionality reduction techniques along with deep learning will soon revolutionize the field of metabolomics.

8.5.2 AI-Based Selection of Bioactive Candidates

A key challenge in the field of phytochemical discovery is how, among so many metabolites produced through untargeted metabolomics, only few merit isolation and characterization. AI approaches are powerful solutions by linking metabolomic features to bioassay readouts, allowing for systematic candidates' prioritization.

Such models will associate, after high-quality peak detection and classification with CNN-based tools, individual or combined features with their bioactivity labels (e.g., cytotoxic vs. inactive extracts) using machine learning models including random forests, gradient boosting, and shallow neural networks to highlight metabolites that most likely account for the observed bioactivity.

These models are further refined by active learning strategies that iteratively select the most informative fractions or features to be tested. In this way, exploration of novel chemical regions is usefully balanced with the exploitation of known structure-activity trends such that experimental resources are used efficiently.

Moreover, semi-supervised approaches can take advantage of the massive amount of unlabeled spectral data to improve the feature representations. This is especially important when the bioassay data are sparse and allows the models to grasp the underlying chemical structure and diversity even in the absence of complete biological annotations.

Taken together, these approaches emphasize how AI can transform metabolite prioritization from a predominantly manual, intuition-driven process into a scalable, data-driven workflow. By integrating spectral analysis with bioactivity prediction, researchers can accelerate the identification of promising phytochemicals while reducing the time and cost of discovery campaigns.

8.5.3 Integration with Chemotaxonomic and Phylogenetic Data

Chemotaxonomy relies upon the observation that similar plants tend to have similar abilities to synthesize certain compounds and that similar compounds are likely to have similar abilities. Phylochemical mapping workflows, which map compound–species associations to the phylogeny, provide evidence that literature-derived data can be used to trace the distribution of lineage-specific compounds and the phylochemical patterns pointing to the evolution of chemical diversity [3].

This idea can be extended to metabolomics. Models using AI will have the ability to use phylogenetic distance, morphological characters, and environmental attributes to predict which lineage will be most closely associated with new metabolites. This brings evolution and metabolomics together to form a model that has more biological predictive ability.

The untargeted metabolomic embeddings and phylogeny of plants can be combined to:

- Find the under-explored lineages: Find the lineages that seem less diverse chemically compared to other related ones, marking these along the way.
- Predict analog production: Estimate whether a species is likely to produce analogs of known metabolites based on their phylogenetic position and metabolomic neighborhood.

Such lineage-resolved predictions supplement traditional data-driven clustering analyses by incorporating the logic of evolution into the metabolomics framework. They thereby enable the directed allocation of resources to evolutionarily informative lineages, thereby speeding up the derivation of novel scaffolds and further integrating plant systematics with chemistry.

These lineage-resolved predictions will therefore supplement traditional data-driven clustering strategies by incorporating the logic of evolution within the metabolomics framework.

8.6 Graph-Based Discovery and Network Inference

8.6.1 Graph Representations in Phytochemical Discovery

The process of phytochemical discovery involves many interacting entities: plants, metabolites, enzymes, pathways, and phenotypes. Together these define the complexity in natural product biosynthesis and function. Graph structures naturally model such interactions where nodes stand for entities and edges capture relationships like "produces," "catalyzes," or "shares biosynthetic precursor."

Knowledge graphs of literature and curated databases, covering relationships among plants, compounds, enzymes, and traits at scale, have enabled the systematic exploration of both biosynthetic logic and ecological associations [3]. Lineage-aware graphs further enable tracing of connections across taxa through conserved biosynthetic modules, thereby underlining underexplored chemical families.

In themselves, at a molecular level, different phytochemicals are graphs of atoms and bonds. Representing molecules as graphs lets AI models directly learn structure–property relationships without handcrafted descriptors. Graph neural networks, along with its related architecture, can easily capture stereochemistry, ring systems, and the connectivity of functional groups, hence yielding predictive power on bioactivity, solubility, and synthetic accessibility.

By integrating knowledge graphs of target biology with molecular graphs of phytochemicals, researchers can construct multi-scale models linking biosynthetic pathways to structural diversity and phenotype. This forms the basis for chemotaxonomic mapping, scaffold prioritization, and AI-driven hypothesis generation that accelerate the discovery of new natural products.

8.6.2 Graph Neural Networks (GNNs) for Compound-Target and Pathway Inference

Graph Neural Networks (GNN) can be considered an extension of neural networks to graph-structured data. GNN applies information flow across graph edges to understand the relationship between related nodes. GNN applications in drug discovery have shown a high degree of accuracy in determining molecular properties and interactions from the graph representation of the target molecules, with higher results compared to many descriptor-based models [18].

For phytochemistry, the GNNs have the following promising applications Prediction of Physiochemical and Pharmacokinetic Properties: GNNs, learning directly from the molecular graphs representing plant compound molecules, can assist in the early-stage triage process of potential metabolites, identifying candidates with desirable solubility, stability, or bioavailability properties.

Hypothesis on biosynthetic pathways A similarity network modeling of metabolites and enzymes helps the GNN hypothesize biosynthetic pathways by revealing the missing enzymatic steps in the pathways.

Completion of knowledge graphs GNNs can predict the missing edges in plant and metabolite knowledge graphs, showing potential, as yet unknown, relations between species and metabolites, as well as between species and traits.

When combined with text-mined phylochemical maps [3], GNNs are able to leverage graph infrastructures both at the molecular level and at the phylogenetic level to predict compound-species pairs. Such lineage-aware knowledge-directed metabolomics allows analytical resources to focus on species with the highest probability of discovering new scaffolds. Through the integration of molecular graphs, biosynthetic paths, and phylochemical maps, GNNs emerge as a framework that is scalable to multi-scale phytochemical discovery.

8.6.3 Mapping Phytochemical and Biosynthetic Networks

Once a substantial list of associations between metabolites, species, and enzymes has been compiled (see Sect. 8.2), network inference approaches can be employed to identify hidden patterns of biosynthetic pathways. For example, a group of structurally related metabolites derived from tyrosine that are abundant in certain plant groups might indicate either convergent or divergent evolution of the corresponding pathways [3]. Additional clues might be gained from the overlap of enzyme annotations and expression networks.

These "graph views" are, however, very different in principle from pharmacological network modeling, as discussed in Chap. 4, and are more concerned with the compound, target, and disease relationships, with an emphasis on human systems. In phytochemistry, the point is more concerned with plant chemical ecology and evolutionary biosynthesis, which utilizes the topology of these networks to focus on the following

1. Enzymes/gene clusters likely to catalyze novel classes of metabolites, informing functional genomics and pathway reconstruction.
2. Species with "bridging" roles that exist between clearly defined chemotypes, and these are believed to possess transitional or hybrid structures, making them valuable taxonomic groups for new chemical discovery.

Through the integration of metabolite-species enzyme relationships and network inference, scientists are now able to progress from the description of chemical diversity toward predictive mapping of biosynthetic potential. Not only does this help expedite the search for new chemical scaffolds but offers evolutionary perspective too, insofar as chemical diversity comes to be seen as related to lineage-specific adaptations and functions.

8.7 Multi-omics Fusion for Lead Identification

8.7.1 Integrating Transcriptomics, Proteomics, and Metabolomics

The genomes of many medicinal plants have not been annotated comprehensively, but the transcriptome, proteome, and metabolome can nevertheless be calculated on a large scale, allowing for vast datasets for discovery. AI-based multi-omics analysis tools allow scientists to predict the association between gene expression, enzymes, and metabolites even in the case of non-annotated pathways.

The capability for learning heterogenous biological data types in bioinformatics has already been established by transformer models and other deep learning models [17]. For phytochemisty, other models will allow one to consider each "omics" level as a view, jointly learning representations emphasizing co-variating modules of genes, enzymes, and metabolites. By this, the hidden associations between the molecular levels will allow one to have a more holistic view of specialized metabolisms.

Through the integration of transcriptomic, proteomic, and metabolomic data, AI models are able to facilitate the interpretation of biosynthetic logic that may not be apparent when single-omics analyses are considered. This will not only aid in understanding pathways and metabolite patterns but will also expedite the identification of novel phytochemicals and their genetic mechanisms.

8.7.2 Identifying Biosynthetic Hotspots and Regulatory Nodes

Integration of multiomics data that helps the models develop using AI provide meaningful and biologically relevant information that helps discover pathways and prioritize metabolites. The primary applications are:

Identification of co-expression modules This involves the identification of groups of biosynthetic genes whose expression patterns are linked with the accumulation of particular families of metabolites.

Detection of nodes of regulation Identifying transcription factors and signaling molecules whose disruption is expected to affect whole branches of specialized metabolism, thus providing targets for metabolic engineering approaches.

Engagement of cryptic pathways Identifying conditions in space, time, or tissue where cryptic biosynthetic pathways become active, opening up potential access to "silent" metabolite spaces.

In the case of plants with not so annotated genomes, and in plants of traditional medicinal significance, the above inferences can serve as important starting material even for the prior studies regarding the function of genes in plants before the comprehensive gene studies for the function of genes in plants are available. In combination with the phylochemical maps" by [3], even the above lineage-aware predictions can aid in prioritizing the plants that have the highest probability of having novel chemistry.

The unification approach makes it possible to connect molecular data with the evolutionary perspective, and it helps to direct the analysis efforts to the hotspots and nodes that regulate chemical diversity by the synthesis process.

The integration and unification technology using AI helps to accelerate the process from data to hypotheses for phytochemical research and reduces the current complexity and limited resolution encountered by traditional approaches and methods.

8.7.3 Case Examples and Practical Considerations

Recent approaches of using large language models and the transformer model paradigm in plant biology have emphasized the need for the integration of diverse datasets, ranging from genomic and phenotypic information to environmental factors, to provide insights into the complex architecture underlying plant traits [19].

Limited omics data When omics data, such as transcriptomics and metabolomics, are limited to certain species, unsupervised and weakly supervised machine learning models are also capable of identifying co-regulated gene and metabolite modules.

Transfer learning Knowledge transfer from model plants to non-model medicinal plants might help address the problem of data scarcity. Nevertheless, domain shift needs to be addressed while performing predictions on a diverse group of species having a different evolutionary background.

There are still some practical issues that need to be addressed. Heterogeneity in data sources, batch effects, and missing data in metadata could potentially mask or reduce valuable biological information that could be better captured by models that are generally more biologically relevant. Standardized pipelines and experimental designs are necessary in order to maximize benefits in the area of AI-assisted multi-omics fusion analysis because only then would multi-omics integration help in rapidly discovering biosynthetic hotspots and regulators in plants even if their genomes are inadequately annotated.

8.8 Challenges in AI-Driven Phytochemical Discovery

8.8.1 Data Sparsity and Heterogeneity

Data banks of plant metabolites are necessarily incomprehensive and lack standardization (Table 8.1). A large number of compounds have been documented only once and sometimes incompletely or ambiguously named. The availability of spectra is greatly biased toward frequent sub-structural motifs; this reduces the scope for (Data Manipulation Language) DMLs to explore the chemical space. In text mining, even with LLM-based pipelines, false negatives are a matter of continued concern: in the context of plant metabolomics, species compound co-occurrence pair extractions have been seen to have a false-negative rate of between 40 and 55% when compared to traditional approaches [6].

The heterogeneity of experimental protocols in the area of metabolomics, NMR, and bioassays introduces domain shifts and threatens the generalization of the models. It has been found that the models trained on the data of a certain instrument or

Table 8.1 Challenges and mitigation strategies in AI-driven phytochemistry

AI approach	Input data	Key applications	Limitations	References
NLP/LLMs	Text (literature, ethnobotany)	Entity extraction and phylochemical maps	False negatives and normalization issues	[1]
CNNs/spectral AI	LC-MS, GC-MS, NMR spectra	Peak detection and functional group prediction	Domain shift and dataset bias	[8]
Generative models (VAEs, GANs, transformers)	Molecular graphs, SMILES	Scaffold hopping and analog generation	Requires curated training sets	[18]
GNNs	Molecular and knowledge graphs	Property prediction and pathway inference	Interpretability challenges	[17]

laboratory may not generalize well on the data of another laboratory, unless special care has been taken to develop robust models or retrain them with the data from the new laboratory [8].

8.8.2 Model Bias, Overfitting: Implications for Interpretability in Complex Systems

Models are bound to learn and reproduce biases present in the data, which will focus more on species/chromotypes that have already been extensively explored and ignore species/chromotypes that are still unexplored. Overfitting could easily happen while training models with smaller sample sets, which could be the case with metabolomics or less-explained species of medicinal plants.

Interpretability is another big concern. Although the applications of attention mechanisms and saliency maps in spectral and text minings are promising for interpreting the results [16], the lack of interpretability makes deep learning imperfect. A recent survey in bioinformatics stresses that model complexity, interpretability, and cost should be balanced [17].

8.8.3 Reproducibility, Benchmarking, and Standardization

Reproducible AI pipelines require open data, documented preprocessing, and accessible code. However, existing models have incomplete pipelines or are based on proprietary tools, which challenge transparency and reproducibility. There have been appeals within NMR and related areas for community benchmarks, datasets, and tools to set the record straight and push the field forward [10].

8.9 Future Directions and Ethical Issues

8.9.1 Democratization of AI Technology for Biodiverse

Democratization of AI in resource-limited areas demands lightweight, open-source software, amenable to execution on low hardware, in addition to training sessions aimed at developing expertise. The existing language and spectral models may be utilized and further trained using data produced within these areas.

Federated learning is promising because models can be jointly trained across multiple institutions without actually centralizing sensitive data, thus facilitating global collaboration while maintaining *data sovereignty* [6]. These models allow biodiversity hotspots to be active contributors to discover pipelines rather than just passive sources of knowledge.

Table 8.2 Ethical and practical considerations for AI in phytochemistry

Dimension	Best practices/principles	Example implementation
Democratization	Lightweight open-source tools and training	Federated learning in biodiversity regions
Indigenous knowledge	FAIR + CARE principles	Access controls and benefit-sharing
Collaboration	Open pipelines and shared datasets	Community-curated phylochemical maps
Standardization	Benchmarks and reproducible workflows	Public spectral libraries and code release

8.9.2 Ethical Bioprospecting and Indigenous Knowledge

The usage of text mining by AI and ethnobotanical intelligence poses serious ethical issues (Table 8.2). The automated usage of traditional information from literature and grey literature can result in an infringement of consent and biopiracy. Recent reviews in plant biology and AI have made it clear that it is necessary to consider quality, legalities, and copyright concerns while compiling a training data set [6].

8.9.3 Open-Source and Collaborative Discovery

For the successful application of AI-phytochemical discovery, collaborative, open Ecosystems are required. Open pipelines and databases based on large-scale text mining and LLM computations have proven highly valuable in the community within material science and plant metabolism studies [3]. For phytochemistry, this means:

A future where shared phylochemical maps, spectral libraries, and multi-omics datasets are being systematically expanded using automated extraction methods. Modular AI tools for text, spectra, graphs, and omics data analysis can be assembled into customizable workflows according to the research question of interest. Crossdisciplinary collaborations between chemists, plant biologists, data experts, and indigenous knowledge-keepers are ubiquitous. Through the integration of technical innovation and ethical and inclusive practices, AI can assist with the transformation of phytochemical discoveries from the current state to become a systematic and globally shared enterprise, with respect to the diversity that this field relies upon.

References

1. Busta, L., et al. (2024). Mapping of specialized metabolite terms onto a plant phylogeny using text mining and large language models. *The Plant Journal, 120*(1), 406–419.
2. Shetty, P., et al. (2023). A general-purpose material property data extraction pipeline from large polymer corpora using natural language processing. *NPJ Computational Materials, 9*(1), 52.
3. Busta, L., et al. (2024). Phylochemical mapping of natural products onto the plant tree of life using text mining and large language models. *bioRxiv*, 2024.02.16.580694.

4. Yang, Y., & Lawson, D. J. (2023). HTRX: An R package for learning non-contiguous haplotypes associated with a phenotype. *Bioinformatics Advances, 3*(1), vbad038.
5. Dagdelen, J., et al. (2024). Structured information extraction from scientific text with large language models. *Nature Communications, 15*(1), 1418.
6. Knapp, R., Johnson, B., & Busta, L. (2025). Advancing plant metabolic research by using large language models to expand databases and extract labeled data. *Applications in Plant Sciences, 13*, e70007.
7. Kantz, E. D., et al. (2019). Deep neural networks for classification of LC-MS spectral peaks. *Analytical Chemistry, 91*(19), 12407–12413.
8. Mitrova, T., & Melnikov, Y. (2019). Energy transition in Russia. *Energy Transitions, 3*(1), 73–80.
9. Zeng, Z., et al. (2024). Cucurbitacin B targets STAT3 to induce ferroptosis in non-small cell lung cancer. *European Journal of Pharmacology, 978*, 176805.
10. Li, M. Y., et al. (2025). Application and research progress on artificial intelligence in the quality of Traditional Chinese Medicine. *Frontiers in Pharmacology, 16*, 1687681.
11. Jahangiri, M., et al. (2022). Early and long-term outcomes of conventional and valve-sparing aortic root replacement. *Heart, 108*(23), 1858–1863.
12. Han, T., et al. (2025). Signatures of chiral superconductivity in rhombohedral graphene. *Nature, 643*(8072), 654–661.
13. Fine, G. A., & Abramson, C. M. (2020). Ethnography in the time of Covid-19. Vectors and the vulnerable. *Etnografia e ricerca qualitativa, 13*(2), 165–174.
14. Barnabe, A., et al. (2025). Convolutional neural networks assisted peak classification in targeted LC-HRMS/MS for equine doping control screening analyses. *Analytical Chemistry, 97*(6), 3236–3241.
15. Reher, R., et al. (2020). A convolutional neural network-based approach for the rapid annotation of molecularly diverse natural products. *Journal of the American Chemical Society, 142*(9), 4114–4120.
16. Wang, Y., et al. (2025). Modeling protein–ligand interactions for drug discovery in the era of deep learning. *Chemical Society Reviews, 54*, 11141.
17. Zhang, D., et al. (2024). Mm-llms: Recent advances in multimodal large language models. *arXiv preprint.* https://arxiv.org/abs/2401.13601
18. Öztürk, E., & Derin, G. (2020). Psikotravmatoloji. *Aydın insan ve toplum dergisi, 6*(2), 181–214.
19. Yoosefzadeh-Najafabadi, M. (2025). From text to traits: Exploring the role of large language models in plant breeding. *Frontiers in Plant Science, 16*, 1583344.

AI-Assisted Development of Personalized Phytopharmacological Therapies

9

Umme Ammara, Rina Agustina, and Qurat ul ain Sajid

Abstract

With increasing interest in phytochemicals as therapeutic agents, efforts are underway to evaluate their efficacy and safety using computational methods. People have been curing with plants for generations, but now for the first time ever, artificial intelligence is improving this process. Plant-based medicines have been around for centuries, but now, AI is allowing scientists to discover new, previously unknown plant natural therapeutics. AI helps us analyze and detect patterns in data that were previously thought to be impossible. Because of this, scientists can spend their time more efficiently on the most promising molecules. AI gives scientists the ability to document and quantify parameters in real time, providing rational and objective therapeutic modifications. AI is in every field of Medicine, including Diagnosis, Therapy Optimization, Patient Management, and the expedited release of New Medicine. In Diagnostics, AI integrates everything past medical history, relevant images, genomics, and more often achieving results with greater accuracy and detail than standard methods. Machine learning improves the management of health of populations, resulting in fewer hospital readmissions and lower overall costs. AI also analyzes genetic and metabolic data to identify higher-risk patients, forecast the course of diseases, and detect the subtle signs that even the most seasoned clinicians can overlook. In this sense, the field of natural medicines is right now undergoing a huge transforma-

U. Ammara
Department of Botany, Rawalpindi Womn University, Rawalpindi, Pakistan

R. Agustina (✉)
Department of Pharmacy, Faculty of Pharmacy, Hasanuddin University, Makassar, Indonesia
e-mail: rinaagustina@unhas.ac.id

Q. u. a. Sajid
Department of Animal Nutrition and Feed Science, Faculty of Biology and Animal Science, Wrocław University of Environmental and Life Sciences, Wrocław, Poland

Y. Bibi et al. (eds.), *Artificial Intelligence in Phytochemistry*,
https://doi.org/10.1007/978-981-92-0875-3_9

tion. Since the discovery of sophisticated tools, including spectroscopy, Natural Medicine has been undergoing significant advancements. Personalized medicines employ the use of genomic sequencing in identifying genes that affect an individual's response to medicines and his or her susceptibility to certain diseases. It is applicable to pharmacogenomics to provide tailored biologic agent dosages to patients. Biomarker analysis is used to determine appropriate targeted therapies as well as predict the progression of disease. Multi-omics analyses, such as proteomics and metabolomics, help provide additional biological insights. Digital health tools and AI process various types of data for comprehensive monitoring and precise treatment. This chapter highlights how AI enables predictive, scalable, and personalized phytochemical drug development.

Keywords

Phytochemicals · Artificial intelligence (AI) · Drug discovery · Personalized medicine · Multi-omics analysis

9.1 Introduction

Pharmacologists are continuously working on interdisciplinary area such as phytochemistry, pharmacognosy, drug development, and pharmacology [1]. For pharmacologists, ethnomedicine and traditional plant therapeutics, in the form of folk medicine, works as a reservoir of therapeutic compounds [2]. Today, the majority of experts have their sights on phytopharmacology as the most promising way for sustainable and personalized medicine [3]. One of the major advantages is the lower side effects when plant-based therapies are offered as an alternative to the synthetic drugs.

The incorporation of artificial intelligence (AI) to the study of natural products has potential to change the paradigm in the search of the next novel therapeutics [4]. The capacity of AI to automatically analyze large datasets and recognize patterns and predict outcomes will streamline the discovery of many compounds that offer varying degrees of therapeutic benefits [5]. One of the ways AI can be applied to natural product research is through virtual screening whereby compounds that reside in large chemical libraries can be screened for a particular biological activity [6]. This strategy will greatly increase the odds of identifying novel therapeutics while simultaneously decreasing the time and cost of the screening process.

AI's usefulness in the healthcare system continues to rise with the ease of supportable decision-making and improvement of clinical outcomes. The increasing volume of daily generated health-related data creates an opportunity for AI predictive models to make immediate informed decisions (Fig. 9.1) [7]. AI's predictive analytics can identify at-risk patients before problems occur [8]. Despite their therapeutic promise, conventional phytochemical drug development remains resource-intensive and experimentally demanding. This has catalyzed a shift toward computational strategies, particularly AI-driven models. AI is, therefore, in the real sense of the word, a resource for the clinicians and public health policymakers striving to remove barriers and make healthcare more accessible and equitable [9]. Tailoring a type of therapy to

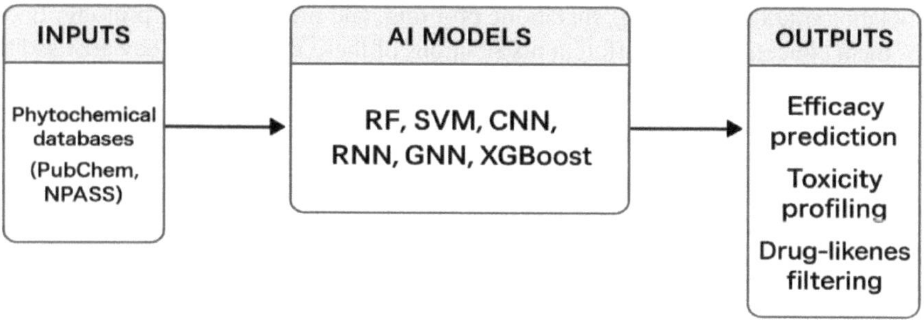

Fig. 9.1 AI workflow for phytochemical drug prediction

an individual is a long-held goal for the healthcare industry. The advancements of AI technology will help with this goal by enhancing the quality of healthcare through analysis of data to help improve the quality of healthcare to assist with the planning of diagnosis and improving treatment options. The healthcare industry is positively impacted by AI through analysis, treatment, improving patient interaction, and enhancing the speed of pharmaceutical development [9].

Predicting the properties in AI such as toxicity, solubility, and pharmacokinetics that machine learning systems use to forecast [10]. This aids in the targeting of new drug development and the recognition of the compounds with the highest success probability. The positive impact of AI on the field of drug development through the use of the molecules that were previously undetermined and accelerate the therapeutic advancements is a powerful force [11]. The increasing number of human beings affected by certain diseases means unexamined potential remains for the examination of natural systems for therapeutic advancements. This potential is vast due to the resources that have been present for millions of years [12]. This methodology entails inspecting natural entities including plants, microorganisms, and sea life for biological activities of interest and qualitatively assessing their commercial transformability to new pharmaceuticals [13].

Using this technology could provide therapeutic activities of interest and address the global public health crises more expeditiously and efficiently. Because of the unprecedented ability of Artificial Intelligence to potentially research and investigate natural products, this study will seek to examine the use of Artificial Intelligence for researched and developed natural products.

9.2 AI-Driven Patient Stratification

9.2.1 Genomic and Metabolomic Profiling for Patient Subgroup Identification

Metabolomics investigates the millions of tiny molecules called metabolites that are always floating and changing in our bodies. It identifies which metabolites are present and their concentrations. Scientists do this in three primary manners:

metabolomic fingerprinting, metabolic profiling, and metabolic foot printing. Each of them concentrates on different distributions of these molecules. They are used to monitor a disease's trajectory in the body, determine recovery rates, or examine responses to surgery or new medications. Each method is used to provide different answers to the puzzle of a disease.[14].

Advanced predictive modeling techniques have changed the way machine learning is used to determine the potential trajectory of a disease in a particular patient. Predictive modeling achieves accuracy rates over 85% [15]. Predictive modeling describes the art of disentangling complex patterns in a set of health parameters. It identifies potential complications in a patient up to 6 months before the complications develop. Predictive modeling works on static and dynamic datasets to establish the flow of the patient. It can evaluate more than 200 clinical parameters simultaneously, increasing overall early detection of chronic diseases in a population by 40% [16].

A desire has developed within these companies to identify even the most nuanced and hard-to-find indicators of underlying disease within the data they analyze. With the incorporation of ML algorithms, population health has reached another level, enabling the appropriate high-dimensional data such as demographics, clinical history, and social history to be stratified into risk categories and realized a 30% reduction in hospital readmissions [17]. A 25%–30% reduction in ED (Emergency Department) overcrowding has been reported, with a $20,000 reduction in healthcare costs per patient per year as well due to the nature of these population health approaches [17]. Automated systems that analyze data in real-time are reported to have an adverse event reduction of 45%, as well as a 50% reduction in critical care bed occupancy, and to have detected certain health events up to 48 h in advance [15].

Clinical decision support systems process large volumes of clinical data in real time, improving diagnostic accuracy by 35% and aligning treatment suggestions with current research [17]. Real-time monitoring systems analyze around 1000 data points per minute, reducing response time to critical events by 40% and improving outcomes by 30% [15]. The integration of these components creates a comprehensive framework for patient risk assessment and stratification, producing a 28% reduction in length of stay and a 32% decrease in preventable readmissions among high-risk patients [17].

9.2.2 AI in Disease Diagnosis

Several AI algorithms have been reported to enhance diagnostic efficiency, especially deep learning algorithms and machine learning [18]. Using data from a patient's history, images, and genetics, as well as other data inputs, AI can find trends and relationships that would not be clear to a clinician. For example, the use of AI in image recognition can assist in identifying slight differences in the image captured through X-rays or MRI for the diagnosis of cancer, heart diseases, and neurological disorders [19]. AI offers a further benefit in the examination and incorporation of the vast volume of data gathered from many sources. Employing artificial intelligence to examine and cross-reference clinical and genetic data can help uncover significant disease-specific biological markers that will improve exact identification and treatment of patients [20]. This leads to avoiding misdiagnoses and increases the chances of treatment intercepting the disease progress at an earlier stage. Research has explored the applicability of AI in various illnesses, such as Alzheimer's, cancer, and cardiovascular [21]. Predictive AI used in the diagnosis of these diseases and disorders has been successful in improving prediction and diagnosis accuracy, and sensitivity and specificity rates, compared to the traditional systems. In the areas of disease diagnosis, disease classification, and medical interventions, AI has been applied and has had a great impact, leading to better choice of treatments and consequently improved patient outcomes. Improving diagnosis of diseases is the core focus of AI. Chen et al. [16] has applied to a collection of works concerning AI in diagnosis, treatment, and prognosis of urologic conditions.

AI is considered to offer greater benefits than traditional systems and practices. The AI superiority gap in diagnostics and predictive healthcare outcomes is in the 71.8% range [9]. Keenan and colleagues studied the ability of retinal specialists to detect retinal fluid in spectral domain OCT scans of the eyes with age-related macular degeneration in comparison to an AI algorithm. The suboptimal accuracy and reduced sensitivity of retinal experts in fluid detection was more evident in the more difficult and less obvious cases of intraretinal fluid: in contrast, the AI systems had more accuracy, more sensitivity, more specificity, and more precise fluid detection in those cases [22].

9.2.3 AI in Personalized Herbal Medicine

The use of AI helps protect the quality and standardization of herbal products and helps create customized strategies for personalized medicine [23]. Quality control and the capacity to tailor treatments based on patients' medical histories, it represents two important challenges in ancient medicine. In terms of quality control, AI is working with spectroscopy and chromatographic control. There are histograms of infrared and the near infrared. Infrared and near infrared spectroscopy combined with AI helps in the automated analysis of control histograms and peak detection [24]. The period of automated analysis, the purity and authenticity of herbal products should be controlled.

AI algorithms defend the safety and reliability of herbal remedies by identifying adulterants and contaminants [25]. AI helps improve versatility and effectiveness by standardizing the chemical profiles of herbal products, which helps improve consumer confidence and meets regulatory requirements. AI is making strides in personalized medicine by developing algorithms that analyze enormous datasets to design tailored treatment plans considering individuals' health data, genetics, and lifestyle. For herbal medicine, AI finds and streamlines personalized approaches to target new drug candidates and maximize treatment efficacy. Moreover, these algorithms create personalized herbal medicine combinations to suit individual patients' needs [26]. This also applies to customization of herbal supplements as demonstrated by the Natural Medicine app. Optimizing treatment through AI offers endless possibilities, the future of natural healthcare lies in AI personalized medicine. These innovations will improve treatment efficacy and safety, drug development will become cheaper, and it will improve accessibility and affordability [27]. Integrating AI technology in relation to quality control and personalized medicine will provide stakeholders in the herbal products industry with the tools needed to navigate complex challenges. To provide the efficacy and safety of traditional herbal remedies in the omnipresent healthcare landscape, AI technology will have to be integrated. Data privacy, accessibility, and regulatory compliance will have to be addressed. These challenges will define the integration of AI technology in herbal medicine [28].

9.3 Real-Time Application

9.3.1 Clinical Evidence of Ashwagandha in Stratified Therapies

There have been randomized and controlled clinical trials of the *Withania somnifera* (ashwagandha) of which results have shown consistent and reproducible physiological effects that relate to both metabolic and stress-response pathways, which would make it a possible candidate for precision (stratified) interventions. As an illustration of this, one randomized study and a larger scale one shortly after it found that after 8 weeks of administering standardized and aqueous root/leaf extracts of *W. somnifera*, adult participants experienced a decrease in validated scores of chronic stresses and positively altered biochemical stress markers providing evidence in support of the herb's modulatory influence on the neuroendocrine and metabolic axes that often intertwine with glycemic control and inflammation. These clinically observed, measurable biomarker changes (e.g., cortisol/stress markers, insulin-related readouts in some trials) provide concrete end-points that could be used as stratification labels in an AI pipeline (i.e., patients with a stress-dominant metabotype vs. those with predominant insulin-resistance signatures).

9.3.2 AI-Stratified Randomized Trials for Enhanced Sensitivity

The impact of AI on the quality of medical studies is evident through the recent review. AI is able to find the optimal patients for studies, strengthen study outcomes, and provide accurate predictive analytics on study outcomes. The use of AI is not limited to the medical and pharmaceutical sectors. The positive effects of AI can also be felt in the chemical and biological engineering disciplines studying plant-based therapies. Lam et al. [29] reviewed a collection of studies documenting the advanced AI capabilities beyond diagnostic. AI can spot relevant trends in study data and aptly select patients for clinical trials. For example, in a clinical trial involving the use of certain herbs to treat diabetes, AI can extract participants who have inflammatory diseases or high oxidative stress. This can provide a positive signal for statistically significant outcomes. Basically, machine learning groups patients more precisely before the trial even kicks off. When the dust settles, you can actually tell what works and what does not.

9.3.3 Proof of Concept-AI Re-analysis Rescues Stratified Subgroup Signals

AI-based proof-of-concept analyses and retrospective re-analyses unlock valuable treatment-response indicators by detecting hidden patient subgroups that traditional statistical techniques fail to capture. Machine learning models like unsupervised clustering, probabilistic topic modeling, and gradient boosted decision trees are capable of finding heterogeneous treatment effects by reallocating patients into more defined strata and then outcomes are analyzed again. Studies demonstrate that subphenotypes derived from AI algorithms reveal subgroups with different risk and treatment benefits profiles. By example, using k-means clustering on latent patient profiles extracted from EHRs illustrates the summarization of the original analysis into visible high- and low-risk subgroups [30], while XGBoost retrospective analyses show that most trial participants were non-responders concealed by the overall positive trial outcomes, which in turn "rescues" the actual responder signal [31]. Similar methodologies in oncology studies demonstrate that AI stratification can identify responder-enriched subgroups and clinically unseen early biomarkers retrospective of the trial ([32]). Within oncology research, AI-assisted data stratification for trial sampling has been shown to demonstrate predictive backward responder sub-group segmentation and early indicator phenomena which traditional/standard retrospective data analyses fail to reveal [33]. All of these investigations demonstrate the potential of AI to uncover clinically relevant, hitherto hidden signals in stratified sub-groups, thereby promoting the design of precision-based trials and the individualized use of phytotherapeutic or pharmacotherapeutic agents.

9.3.4 Integrated Omics and Herbal Knowledge Platforms

Currently, an expanding array of artificial intelligence platforms, including databases for biological data, chemical data, and compounds, are being united (Fig. 9.2). This integration will allow true personalized phytotherapy, tailored to align with an individual's exact requirements. Industrial and academic sectors are integrating multiomic with clinical data to identify optimal targets and compounds. This even allows research teams to conduct virtual screenings of entire chemical libraries. Comparison of databases containing phytochemicals to those containing ADME/Tox prediction tools alongside a patient's molecular profile enables the accurate determination of the most suitable botanical therapy [34].

Overall, most databases and tools focus on small-molecule drugs (Table 9.1). There is little to no difference in incorporating the phytochemicals, as the same approach is applicable. Now, this is combined with AI-operated clinical trials, and personalized herbal medicine is in the near future.

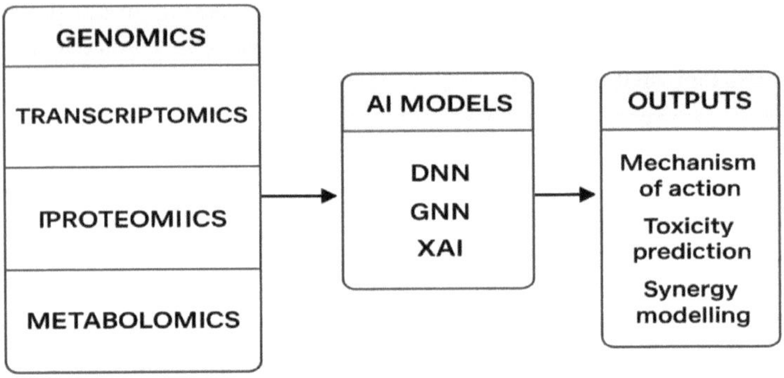

Fig. 9.2 Multi-omics integration framework

Table 9.1 AI model comparison table

Model	Accuracy	Interpretability	Scalability	Data requirement	Use case versatility
Random Forest (RF)	Moderate–high	High	Moderate	Low–medium	Good for baseline efficacy/toxicity tasks
Support vector machine (SVM)	High on well-separated classes	Medium	Low–moderate	Medium	Effective for binary classification, drug-likeness filtering
XGBoost	High	Medium–high	High	Medium	Versatile for efficacy prediction, toxicity profiling
Convolutional neural network (CNN)	High	Low–medium	High	High	Good for sequence-based tasks, pattern recognition
Recurrent neural network (RNN)	Moderate–high	Low	Moderate	High	Useful for SMILES generation, sequential toxicity signals
Graph neural network (GNN)	Very high	Medium	High	High	Best for structure–activity prediction, scaffold generalization

9.4 Global Case Studies

9.4.1 Case Study 1: AI-Driven Herbal Matching in Europe

Recent achievements in artificial intelligence have impacted drug repurposing. Predictive systems use biomedical data such as clinical symptoms, drugs, as well as genes with their orthologs to improve predictions. Deep learning systems assimilate and evaluate multi-omics data and drug data to forecast undocumented actions of drugs and identify new clinical uses. Some such systems map drug-target interaction networks, which enhances repurposing prospects. These systems can also identify and flag candidates that may have significant safety issues for repurposing and are being improved for use in routine clinical practice. The technology is currently assisting in the identification of novel drug candidates for some rare diseases and complex cancers, for example, glioblastoma [35]. Most notably, such novel approaches to drug development capture new uses of existing drugs, streamline the development of new uses, and lower costs [36].

9.4.2 Case Study 2: AI in Herbal Safety and Standardization

Siqueira et al. [37] discuss the implications of the intersection of AI and the study of traditional medicine and the implications of this intersection. While this field looks through the lens of AI, traditional medicine and the study of it look through the lens of traditional medicine. The focus of this hawk-eye approach remains on autonomy, beneficence, and the principles of non-maleficence and justice. In the range of provided beneficence, the non-maleficence of complex changes is withdrawn through privacy and consent. Framed by the principle of non-maleficence with respect to privacy, healthcare governance should balance risk and autonomy to ensure that vulnerable patients and engaged participants are not disproportionately exposed to harm. David et al. [38] proclaim that complexities of governing healthcare AI demand regulations that the EMA and the FDA lack; in simple terms, without adequate regulations, AI technology in healthcare is not acceptable. Thus, the sensitive nature of health information analyzed by AI poses the implications of privacy regulations, informed consent, and the standards of data collection as ever more necessary.

Algorithmic opacity is proposed as a regulatory and accountability mechanism as it is the mechanism designed to explain how algorithmic decisions are made primarily to safeguard the autonomy and rights of individuals [39]. Real-time instances involve the utilization of AI to examine genetic data for potential personalized herbal therapies, raising issues of genetic data privacy and consent, and regulatory agencies requiring AI in drug discovery for safety and efficacy to be validated and documented [40]. WHO data confirms 30% of nations have in place regulatory structures concerning traditional and complementary medicine, illustrating an international interest in the safety, efficacy, and ethical concerns surrounding the research of herbal medicine. The incorporation of AI in the research of herbal medicine offers an opportunity to achieve a truly novel approach, not yet experienced in the balance between pioneering advancements and the ethical constraints of such an undertaking, to protect the wellbeing of patients and the ethical principles of openness [41].

9.4.3 Case Study 3: AI-Powered Formulation Research in India

A research project at the K.B.H.S.S. Trust's Institute of Pharmacy and Research Centre in Nashik, Maharashtra, employed artificial intelligence to expedite the phytochemical screening and the formulation design process for herbal remedies. The target constituents of the herbal remedies were potential anti-inflammatory and antimicrobial compounds. The researchers focused on complex plant datasets and employed machine-learning and data mining techniques to prioritize phytochemicals. Also, natural language-processing tools were used to examine literature and ethnobotanical records to broaden the search and point out useful plant combinations. These AI methods sped up the process, helped prioritize lead compounds, and guided better formulation plans for preclinical testing.

9.5 Smart Delivery System

In an article published in RSC Pharmaceutics in 2025, Panchpuri, Painuli, and Kumar investigate the application of artificial intelligence (AI) in developing intelligent drug delivery systems. Personalized medicine has advanced as a result of this effort. The paper describes how AI modelling and improvement can aid in the development of nanocarriers that regulate medication delivery according to the requirements of specific patients. Better medication action and fewer adverse effects could result from this. The AI concepts could be used to plant-based medications, even though the research focusses on general pharmaceutical applications. Plant-based medicines may be delivered more safely and precisely if nanocarriers were designed to respond to a patient's pain levels or biomarkers [42].

9.6 Emerging Trends

Currently, artificial intelligence is sharpening focus on phytopharmacology. The fusion of ethnobotanical knowledge, wearable technologies, and real-time body tracking and the generative AI modelling of plant compound interactions is a trifecta.

Potential new trends from the synthesis of ethnobotanical knowledge and generative AI modelling from plant compounds offers a new frontier. The data integration from genetic, molecular, and clinical data is a paradigm shift improving the efficacy of predictive modelling and personalized treatment the data integration from different domains to pre-determined outcomes allows the researcher to isolate and draw assumptions from data from different disciplines and focus on the unexplained outcomes. The data sets line to focus on the unexplained alignment of the clinical and biological outcomes and the herbal component to the patient's outcome enhancing the precision medicine opportunities. Innovation and advanced AI and big data techniques emerge from integration of collaborative, technological advancements, data synthesis, and algorithms [40]. The integration of fragmented data sets to advance the empirical knowledge base and evidence-based decision making is enabled.

9.6.1 AI and Ethnobotany: Where Tradition Meets Modern Medicine

AI brings together plant wisdom and modern medicine for a unique interdisciplinary approach. It goes far beyond number-crunching. It employs old manuscripts and folk stories, analyzes them to find plant compounds with unmedicinal potential, and predicts their possible functionality as medicine.

Machine Learning Algorithms

Machine learning algorithms-deep learning convolutional neural network CNN is a deep feedforward neural network with the characteristics of local connection and weight-sharing that uses a stack of convolution layers to extract features. It is widely used in image classification, facial recognition, semantic segmentation [43], object detection [44], and natural language processing [45]. The core idea of CNN involves a local receptive field, weight-sharing, and a pooling layer. The architecture of CNN consists of a sequence of layers that function as follows: when data is input into the CNN, the convolution layer extracts the features, and the pooling layer aggregates the local features extracted by the convolution layer to obtain global features.

Size 1×1, 3×3, or 5×5 convolution filters (also referred to as kernels) are applied in convolution layers. An activation function follows after every convolution layer. The activation function that is often used is the ReLU function, which is defined mathematically as $f(z) = \max(0, z)$, representing the max between z and 0. Afterwards, the pooling layer is applied to lessen the reliance on specific positions, decrease the overall parameters and computation within the network, and help control overfitting [46]. The most common function used for pooling is the MAX pooling function that forms a single neuron in the subsequent layer by taking the highest activation from a defined group of neurons from the previous layer. Average pooling is also a possible option [47]. The pooling layer reduces the dimensionality of the previous layer's input. In doing so, the model becomes more efficient, the computations require less time, the feature maps become more generalizable, and overfitting can be assisted. Three hyperparameters define the amount of output data: depth, stride, and padding. The depth matches the number of filters applied, and the stride parameter indicates how many pixels forward the filter shifts after each slide. The padding layer reflects how the borders of the volume data can be padded with zeros or mean values. The CNN reduces the losses through iterative refinements of network parameters and, as a result, achieves better accuracy through iterative trainings [48].

9.6.2 Natural Language Processing

Taking unstructured data from medical texts and extracting valuable information is one area within healthcare AI that is rapidly improving and becoming increasingly important. Streamlined NLP systems are able to process and analyze clinical notes, literature, and survey response texts with unmatched efficiency (A Survey of Surveys (NLP & ML). There has been a significant improvement with new transformer-based models that are able to comprehend and analyze medical jargon and context without error up to 40% better than older models. Incorporating information from varied sources is one of the major paradigm shifts within healthcare AI and infrastructure [49]. Modern health systems are now able to fully integrate electronic health EHRs, genetic data, lifestyle information, and health data from wearable devices in real time. This broad data integration enables the generation of very granular patient profiles resulting in better outcomes to 35% in predicting the progression of chronic diseases when multiple data sources are applied.

There is a rise in the need for the monitoring of patients' real-time health data for monitoring of health status and risk evaluation. Health trackers today produce over 1000 data points every second, every patient, creating a copious amount of data for monitoring [49]. If this data is integrated with past medical history, real-time insights can reveal deteriorations in a patient's health even 48 hours before changes can be seen with traditional monitoring. On top of that, AI brings everything together genomics, transcriptomics, metabolomics to back up historical claims and show how these treatments actually work at a molecular level [50].

9.6.3 Wearable Tech Integration: Real-Time Monitoring of Herbal Therapies

AI was combined with wearable technology and smartphone apps to monitor the reactions of patients taking *Bacopa monnieri*-based cognitive supplements. The system tracked sleep, mood, and memory variables to feed data into predictive models that modified dose recommendations in real-time. It is a smart health monitoring system that uses wearable technology (like smartwatches) and smartphone apps to track how people respond to an herbal supplement made from *Bacopa monnieri*, a plant thought to improve memory and focus. The following are significant indicators that the system tracks: sleep patterns (the length and quality of an individual's sleep), mood, or emotional state; the ability to recall information is known as memory performance. All of this data is analyzed by artificial intelligence (AI).

The AI learns how each person responds to the supplement by searching for trends. Based on this, it may forecast and modify the herbal supplements dosage in real time; that is, it suggests raising, lowering, or keeping the dosage in accordance with the individual's body and mental state [51]. Major global health issues including diabetes, hypertension, and cardiovascular illnesses put a burden on healthcare systems since they require ongoing, individualized care. Real-time vital sign tracking and predictive analysis using machine learning algorithms are made possible by AI-based health monitoring systems that use smartwatches and mobile applications [52]. AI-assisted monitoring improved patient outcomes, such as lower blood pressure and HbA1c levels and better medication adherence, in a randomized controlled trial that used quantitative and qualitative methodologies [53].

9.6.4 Generative AI in Formulation: Simulating Compound Interactions

The ability of generative AI to generate new ideas is significantly changing our understanding of herbal medicine and the research of plant-based medications. Before we even begin testing them in the lab, we can now utilize computers to obtain a clear understanding of how various natural compounds will interact at a microlevel and even modify their structures. Scientists are using sophisticated computer systems, such as GANs and VAEs, to quickly develop, modify, and test a large number of potential plant-based therapeutic candidates. This expedites the initial stages of discovering novel medications [54]. For instance, VGAN-DTI uses 2D pictures to predict how well a medication would adhere to its target. The effectiveness of novel structures' binding is then assessed using machine learning (Nature Scientific Reports 2025). Additionally, VAEs are utilized to improve molecules in certain ways, such as improving their dissolution, stability, and ease of absorption by the body. This ensures that the potential plant-based medications will function more like pharmaceuticals while remaining natural [55]. Even more advanced systems exist. For instance, X-LoRA-Gemma uses biological principles to create networks that use ongoing feedback and cooperation to help molecules change in ways that could help treat illnesses [56]. All things considered, our knowledge of plant-based medicines is being profoundly changed by AI's capacity to create new things. It is helping us test plant chemicals with computers, which lowers the amount of money we have to spend on experiments when trying to turn intriguing natural substances into medicines that people can actually use.

9.7 Conclusion

In conclusion, the ability to accurately predict efficacy, conduct in-depth toxicity evaluation, and measure drug likeness have advanced the integration of AI into phytochemical research. The ability to assess the impact of deep learning and machine learning on the discovery and evaluation safety processes will only improve over time. Further, the ability to train predictive AI models in a distributed manner will improve transparency and confidence in the models. The full potential of these models will come from the collaboration of chemists, biologists, computational scientists, and clinicians within a well-defined regulatory framework. The impact of these models will go beyond predictive accuracy to the discovery of safer and more effective therapeutics.

References

1. Litjens, G., et al. (2019). State-of-the-art deep learning in cardiovascular image analysis. *JACC: Cardiovascular Imaging, 12*(8 Part 1), 1549–1565.
2. Collaco, B. G., et al. (2025). The role of agentic artificial intelligence in healthcare: A systematic review. *NPJ Digital Medicine*. https://doi.org/10.1038/s41746-026-02517-5
3. Patel, D., & Kore, S. A. (2020). Artificial intelligence: Future impacts, challenges and recommendations on healthcare services. *International Journal of Community Medicine and Public Health, 7*, 1596–1598.
4. Thomford, N. E., et al. (2018). Natural products for drug discovery in the 21st century: Innovations for novel drug discovery. *International Journal of Molecular Sciences, 19*(6), 1578.
5. Williams, A. M., et al. (2018). Artificial intelligence, physiological genomics, and precision medicine. *Physiological Genomics, 50*(4), 237–243.
6. Haick, H., & Tang, N. (2021). Artificial intelligence in medical sensors for clinical decisions. *ACS Nano, 15*(3), 3557–3567.
7. Edison, G. (2023). Transforming medical decision-making: A comprehensive review of AI's impact on diagnostics and treatment. *BULLET: Jurnal Multidisiplin Ilmu, 2*(4), 1121–1133.
8. Fadhel, M. A., et al. (2024). Navigating the metaverse: Unraveling the impact of artificial intelligence—A comprehensive review and gap analysis. *Artificial Intelligence Review, 57*(10), 264.
9. Alum, E. U., & Ugwu, O. P.-C. (2025). Artificial intelligence in personalized medicine: Transforming diagnosis and treatment. *Discover Applied Sciences, 7*(3), 193.
10. Verma, A. K., et al. (2025). Pharmacological approaches and innovative strategies for individualized patient care. *Recent Patents on Biotechnology, 20*(1), 89–107.
11. Alshdoukhi, I. (2022). Review of application of artificial intelligence in natural products research. *Journal of Health Informatics in Developing Countries, 16*(2), 1–11.
12. De Luca, V., et al. (2012). Mining the biodiversity of plants: A revolution in the making. *Science, 336*(6089), 1658–1661.
13. Vrenken, H., et al. (2021). Opportunities for understanding MS mechanisms and progression with MRI using large-scale data sharing and artificial intelligence. *Neurology, 97*(21), 989–999.
14. Barderas, M. G., et al. (2011). Metabolomic profiling for identification of novel potential biomarkers in cardiovascular diseases. *BioMed Research International, 2011*(1), 790132.
15. Swamydoss, D. (2022). Machine learning based predictive analysis of diseases in health care. In *2022 Third international conference on smart technologies in computing, electrical and electronics (ICSTCEE)*. IEEE.
16. Chen, Z. H., et al. (2021). Artificial intelligence for assisting cancer diagnosis and treatment in the era of precision medicine. *Cancer Communications, 41*(11), 1100–1115.
17. Begum, B., Kannimuthu, S., & Hussain, D. M. (2023). An in-depth analysis of automated health monitoring systems for elderly care. In *2023 second international conference on augmented intelligence and sustainable systems (ICAISS)*. IEEE.
18. Nissar, I., et al. (2024). An intelligent and automated machine learning-based approach for heart disease prediction and personalized care. In *International conference on advances in distributed computing and machine learning*. Springer.

19. Alowais, S. A., et al. (2023). Revolutionizing healthcare: The role of artificial intelligence in clinical practice. *BMC Medical Education, 23*(1), 689.
20. Karalis, V. D. (2024). The integration of artificial intelligence into clinical practice. *Applied Biosciences, 3*(1), 14–44.
21. Sharma, V., et al. (2023). Artificial intelligence (AI) integration in medical education: A pan-India cross-sectional observation of acceptance and understanding among students. *Scripta Medica, 54*(4), 343–352.
22. Delanerolle, G., et al. (2021). Artificial intelligence: A rapid case for advancement in the personalization of gynaecology/obstetric and mental health care. *Women's Health, 17*, 17455065211018111.
23. Srija, K., et al. (2021). Artificial intelligence in personalized medicine. In *Artificial intelligence and machine learning in healthcare* (pp. 57–69). Springer.
24. Păcuraru, I.-M., Chirvase, C.-S., & Tiriteu, Ş.-I. (2025). The role of artificial intelligence in personalised medicine: Advancements, challenges, and future perspectives. *Business Excellence & Management, 15*(1), 59–84.
25. Paul, D., et al. (2021). Artificial intelligence in drug discovery and development, *Drug Discovery Today, 26*(1), 80–93.
26. Liu, C., & Ye, F. (2025). A review of multimodal medical data fusion techniques for personalized medicine. In *Proceedings of the 4th international conference on biomedical and intelligent systems*. ACM.
27. Li, Y.-H., et al. (2024). Innovation and challenges of artificial intelligence technology in personalized healthcare. *Scientific Reports, 14*(1), 18994.
28. Abdullah, M. (2024). Integrating artificial intelligence in precision medicine: Enhancing predictive accuracy and personalized Care in Chronic Disease Management. *Medical and Life Sciences, 3*(1), 6–15.
29. Lam, N. F. D., Cai, J., & Ng, K. H. (2025). Artificial intelligence and its potential integration with the clinical practice of diagnostic imaging medical physicists: A review. *Physical and Engineering Sciences in Medicine, 48*(2), 1–16.
30. Subbaswamy, A., & Saria, S. (2020). From development to deployment: Dataset shift, causality, and shift-stable models in health AI. *Biostatistics, 21*(2), 345–352.
31. Wang, D., & Zhang, S. (2024). Large language models in medical and healthcare fields: Applications, advances, and challenges. *Artificial Intelligence Review, 57*(11), 299.
32. Dharani, S., & Kamaraj, R. (2024). A review of the regulatory challenges of personalized medicine. *Cureus, 16*(8), e67891.
33. Sharma, A., et al. (2025). Revolutionizing patient care: A personalized medicine approach using artificial intelligence for better healthcare. In *AI-driven breakthroughs in antimicrobial resistance* (pp. 169–194). IGI Global Scientific Publishing.
34. Saripalli, H. R., et al. (2025). *Journal of Population Therapeutics & Clinical Pharmacology*. Mikky Publications.
35. Dagdelen, J., et al. (2024). Structured information extraction from scientific text with large language models. *Nature Communications, 15*(1), 1418.
36. Xuyan, H., et al. (2025). Visual-language reasoning large language models for primary care: Advancing clinical decision support through multimodal AI: X. Huang et al. *The Visual Computer, 41*, 1–22.
37. Siqueira, A. D. S. E., et al. (2025). *Artificial intelligence in cancer control actions: Solution or problem?* (p. e-005291). SciELO Brasil.
38. David, E., et al. (2024). Thyroid nodule characterization: Overview and state of the art of diagnosis with recent developments, from imaging to molecular diagnosis and artificial intelligence. *Biomedicine, 12*(8), 1676.
39. Kurzawinski, T. R., Butler, C. R., & Aziz, T. A. (2025). MEN2: Surgical precision in the era of precision medicine. *Endocrine-Related Cancer, 32*(6), e240251.
40. Azadnia, R., et al. (2022). An AI based approach for medicinal plant identification using deep CNN based on global average pooling. *Agronomy, 12*(11), 2723.

41. Alonazi, R. (2025). Leveraging AI to identify medicinal herbal plants properties. In *Transforming pharmaceutical research with artificial intelligence* (pp. 247–274). IGI Global Scientific Publishing.
42. Prusty, A., & Panda, S. K. (2024). The revolutionary role of artificial intelligence (AI) in pharmaceutical sciences. *Indian Journal Of Pharmaceutical Education And Research, 58*(3s), s768–s776.
43. Chowdhury, M. E., et al. (2021). Using deep learning technique. In *Technology in agriculture* (p. 453). InTechOpen.
44. Zaharia, O. P., Schrauwen-Hinderling, V. B., & Roden, M. (2022). Imaging in precision medicine for diabetes. In *Precision medicine in diabetes: A Multidisciplinary approach to an emerging paradigm* (pp. 89–110). Springer.
45. Schork, N. J. (2015). Personalized medicine: Time for one-person trials. *Nature, 520*(7549), 609–611.
46. Soffer, S., et al. (2019). Convolutional neural networks for radiologic images: A radiologist's guide. *Radiology, 290*(3), 590–606.
47. Ben-Cohen, G., et al. (2022). TP53_PROF: A machine learning model to predict impact of missense mutations in TP53. *Briefings in Bioinformatics, 23*(2), 1–19.
48. Wang, S., et al. (2019). Pathology image analysis using segmentation deep learning algorithms. *The American Journal of Pathology, 189*(9), 1686–1698.
49. Tsumoto, S., & Hirano, S. (2010). Risk mining in medicine: Application of data mining to medical risk management. *Fundamenta Informaticae, 98*(1), 107–121.
50. Ashraf, H., Alam, A., & Ahmad, R. (2024). Personalised medicine. In *Personalized and precision nanomedicine for cancer treatment* (pp. 5–13). Springer.
51. Karmakar, D., Bora, D., & Keot, D. (2025). Indian spices in sustainable healthcare: A review. *The Agbiz Today, 2*(4), 19–28.
52. Rajkomar, A., Dean, J., & Kohane, I. (2019). Machine learning in medicine. *New England Journal of Medicine, 380*(14), 1347–1358.
53. Mackey, T., et al. (2020). Machine learning to detect self-reporting of symptoms, testing access, and recovery associated with COVID-19 on twitter: Retrospective big data infoveillance study. *JMIR Public Health and Surveillance, 6*(2), e19509.
54. Afsar, Z., et al. (2025). Modern concepts of integrating artificial intelligence with traditional and herbal medicine: A review. *International Journal of Scientific Research and Technology, 2*(2), 43–49.
55. Umer, M. S., et al. (2025). Large language models meet molecules: A systematic review of advances and challenges in AI-driven cheminformatics. *Archives of Computational Methods in Engineering, 32*(7), 1–42.
56. Stewart, I., & Buehler, M. J. (2025). Molecular analysis and design using generative artificial intelligence via multi-agent modeling. *Molecular Systems Design & Engineering, 10*(4), 314–337.

Collaborative Strategies for Pharma Companies Using AI in Phytopharmacology

10

Iram Zaheer, Tayyaba Afzal, and Jaroslaw Prockow

Abstract

The healthcare industry is being completely reshaped by the technological revolution of "artificial intelligence." For phytopharmacological research, artificial intelligence is doing more than just speeding up the processes of extraction and bioassay testing but is also completely transforming the whole approach. Currently, drug discovery is a kind of team sport, which is led by the data science. The pharmaceutical industry has partnered with AI-focused organizations. Open-source platforms are becoming increasingly popular, since scientists are trying to find a balance between protecting new concepts and scientific progress. There is a continued effort to define a framework for AI in phytopharmacological research, which is focused on a technological foundation and teaming up with scientists for trustworthy outcome achievements. The integration of artificial intelligence in phytopharmacological research offers immense possibilities but also demands a responsible application for meaning full growth in health industry.

I. Zaheer (✉)
Department of Environmental Sciences, Fatima Jinnah Women University, Rawalpindi, Pakistan

T. Afzal
Department of Animal Nutrition and Feed Science, Faculty of Biology and Animal Science, Wrocław, Poland

J. Prockow
Institute of Biology, Faculty of Biology and Animal Science, University of Environmental and Life Sciences, Wrocław, Poland

Y. Bibi et al. (eds.), *Artificial Intelligence in Phytochemistry*,
https://doi.org/10.1007/978-981-92-0875-3_10

Keywords

Extraction · Open-source · Computational frameworks · Technological foundation · Responsible AI application

10.1 Introduction

AI has changed the methodologies that were traditionally followed by phytopharmacological researchers. Now we are able to identify intricate patterns, anticipate the behavior of compounds, and optimize them by using less resources and time [1]. Phytopharmacology is the study of bioactive agents found in plants and their mode of action. According to an estimate, a quarter of all pharmaceutical agents that are developed in the past four decades either come from plants themselves or are derived from their derivatives [2].

This chapter is focused on industries, academic labs, and startups that are incorporating artificial intelligence and open science communities, which leads to the advancement toward drug discovery process. The integration of AI in phytopharmacology is more than an upgrade; it is a paradigm shift. Presently, machine learning, modeling, and multispectral databases are all intertwined. This has enabled faster development and discovery of new drugs from plants [3, 4]. The tools have enabled scientists to identify paths, to predict a phytochemical's interaction affinity, and to suggest modifications that can enhance bioavailability or efficacy.

The pharmaceutical industry is also relying on collaborative environments, such as open-data partnerships, scientific knowledge, industrial research, and AI startups. Such environments not only help in the transmission of information, but they also help in minimizing risk, costs, and the exploration of the field of chemical and biological research [5]. The following sections will outline the development of these environments and illustrate cases to analyze their effectiveness.

10.2 The Evolution of Collaboration in AI-Driven Pharmacology

10.2.1 Historical Context: From Data-Driven Discovery to Empirical Screening

Empirical extraction and bioassay-guided isolation are two labor-intensive, low-throughput approaches traditionally utilized in natural product research. Cheminformatics and quantitative structure–activity relationship (QSAR) models were developed during the 1990s digital revolution. By the 2010s, machine-learning techniques based on spectral and docking data were developed. These tools can predict and model molecular interactions of phytochemicals [6].

The pharmaceutical industries have also started embracing open innovation strategies. Large firms, that is, Bayer, Merck, and GlaxoSmithKline, have understood

Table 10.1 Types of collaboration in AI into phytochemical knowledge systems

Sr. no	Collaboration types	Description
1.	Corporate-Startup Alliances	Pharmaceutical corporations collaborate with AI biotech businesses to provide algorithms and data scientists
2.	Public-Private Consortia	Federated or shared-data programs, such as MELLODDY and open targets, offer secure model training across several institutions
3.	Academic-Industrial Symbiosis	Industry provides compound libraries and clinical pipelines, while universities contribute mechanistic expertise and validation models.
4.	Community-Sourced Databases	Initiatives like NPASS and SuperNatural II provide curated phytochemical data

that by forming alliances with laboratories and AI start-ups, they can speed up drug discovery process [7]. As data is keep growing and more computational tools are developing, there is no single institution that has expertise, infrastructure, and data volume that is required (Table 10.1).

10.2.2 Conceptual Change: Integrating AI into Phytochemical Knowledge Systems

Artificial intelligence has significantly transformed the approach and philosophy of phytopharmacology. Previously transmitted through ethnobotanical observation, information is now digitally recorded in structured ontologies that artificial intelligence can mine. Graph neural networks connect plant species, metabolites, and disease pathways, while natural language processing (NLP) technologies scan ethnomedicine literature [8].

When put together, these networks demonstrate how dynamic knowledge ecosystems have replaced linear discovery chains (Fig. 10.1).

10.2.3 Pharmaceutical Companies Collaboration

Pharmaceutical companies pursue AI collaboration for a variety of interrelated reasons:

- According to Zheng et al. [9], AI can improve efficiency and cost-effectiveness by reducing the necessity for wet-lab testing to identify potential leads.
- Machine learning uncovers millions of phytochemical alternatives beyond traditional synthesis, broadening the chemical space.
- Partnerships can mitigate financial and technological risks in early discovery [7].
- Shared governance frameworks ensure transparency in data use and regulatory compliance for AI-generated candidates.

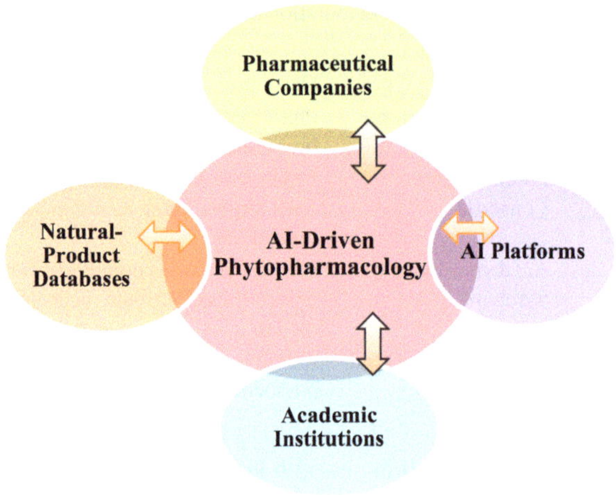

Fig. 10.1 Four nodes: Pharma companies, AI platforms, academic institutions, and natural product databases; arrows showing bidirectional data and knowledge flow, and highlights showing cocreation rather than linear pipelines

These strategic motives explain the rise of cooperation, such as Merck-Exscientia and Bayer-Atomwise, which were discussed in Sect. 11.3.

10.2.4 From Cooperation to Co-creation

The most complex interactions have already progressed beyond simple outsourcing to co-creation, in which AI developers and biopharma researchers work together to construct models and validate experimental results. For instance, Merck works with Exscientia to connect algorithmic design with the company's in-house wet-lab testing; this alliance guarantees iterative feedback between digital prediction and empirical validation [10]. This approach is essential in phytopharmacology, as complex phytochemicals and synergistic effects are challenging for single-target paradigms. AI can provide hypotheses for researchers to refine results based on contextual biological understanding, combining computational and empirical intelligence.

10.3 Case Studies on Strategic AI Partnerships in Drug Discovery

Recent advancements in artificial intelligence (AI) are reshaping drug discovery by enabling collaborations between pharmaceutical companies and computational research organizations. The following case studies illustrate how AI-driven collaborations contribute to drug discovery while offering transferable models for phytopharmacological research in low- and middle-income research settings.

10.3.1 Merck KGaA and Exscientia: Generative AI for Molecular Optimization

In 2023, Merck KGaA entered a multi-program collaboration with Exscientia for development of small-molecule therapeutics in oncology, neuroinflammation, and immunology. This partnership has integrated experimental biology with generative AI models [11].

(a) The platform from Exscientia generates innovative molecular candidates for lead discovery, not relying exclusively on predefined libraries. It is for this reason that multiple parameters like biological activity and pharmacokinetic profile have been simultaneously optimized using this technology [5].
(b) From a Global South perspective, the following tools can be applied for phytochemicals, as some phytochemicals have been discovered, but there is no optimization process for such phytochemicals to be used in clinical trials [1].
(c) This Merck–Exscientia model, in this bidirectional workflow, the findings from experiments help to direct the predictions, and the predictions also help to direct the experiments. This type of workflow would also be beneficial to phytopharmacological research teams with limited access to the laboratory.

10.3.2 Atomwise and Bayer: Deep Learning for Target Prediction

The collaboration between Atomwise and Bayer AG is another example of such collaboration. These parties are applying deep learning approaches in large-scale screening projects. AtomNet produced by Atomwise uses a type of convolutional neural network that uses three-dimensional representations to forecast the interaction between chemical compounds [12].

In this collaboration, Bayer assesses hundreds of millions of compounds to identify a lead compound for making drugs for metabolic and cardiovascular diseases. This process decreases both price and time.

Such an approach would be beneficial for phytopharmacology and more needed in the Global South setting. Phytochemical diversity is extensive, but it would be impractical or impossible to accommodate it in a laboratory setting. There are many plant-derived compounds that have known structures that are not extensively studied regarding pharmacology. Using AI in targeting prediction, one can select the projected best candidate for analysis within a laboratory setting [13].

AI models, developed through huge amounts of synthetically developed data, could be used for smaller data sets related to phytochemicals. This would allow scientists to enable drug discovery efforts to be more inclusive, reaching beyond chemical space that receives limited attention.

Table 10.2 Comparative summary of AI–pharma and AI–phytochemical partnerships

Partner organizations	AI approach	Relevance for global South phytopharmacology	Key contribution	Year
Merck KGaA Exscientia	Generative AI, iterative learning	Optimization of bioactive plant compounds	Multi-target molecular design	2023
Bayer – Atomwise	Deep learning (CNNs), virtual screening	Target prioritization for phytochemicals	Large-scale computational screening	2023
Insilico Medicine – CAS	Reinforcement learning	Modernization of traditional medicine compounds	Improved efficacy prediction	2024
Auransa Inc.	AI-based transcriptomics	Functional validation of plant-derived compounds	Pathway-linked compound discovery	2024

10.3.3 Emerging AI Applications in Natural Product Research

There are a few more projects that are using AI for drug discovery, such as Insilico Medicine and Chinese Academy of Sciences partnership. They are using reinforcement learning to search for compounds derived from traditional Chinese medicine [12].

Auransa Inc. is doing something similar in terms of its use of AI in transcriptomics analysis. They identify plant compounds that align or match gene patterns of specific diseases. Thus, it is less on how similar they look, and it has more to do with their functional activities. That would be very valuable for some tricky areas like cancer and brain-related issues [13].

Such endeavors suggest there is some shift here, where AI is incorporating data not only from artificial sources but also natural sources (Table 10.2). It definitely seems like some kind of bigger shift is going on. For scientists in the Global South, it might well be easier for them now to develop ways in which their own plant knowledge could be put into better pharmacology research, for example. Not everything is quite clear here.

10.4 Data-Sharing Models and Open-Source Platforms

Open science platforms differ in purpose; some focus on targets, others on privacy-preserving model training, and others on phytochemical curation. Academics may share ideas while protecting data security and intellectual property thanks to open data and federated learning initiatives. Because data in phytopharmacology is scattered throughout ethnobotanical archives, regional repositories, and small-scale research databases, these frameworks are essential.

10.4.1 Open Targets: AI-Powered Target Recognition

GlaxoSmithKline (GSK), Sanofi, and other companies are part of the public–private partnership Open Targets, which is led by the European Bioinformatics Institute

(EMBL-EBI) [12]. It combines transcriptomic, chemical, and genomic data to identify and validate treatment targets for a variety of diseases.

Open Targets is a helpful paradigm in phytopharmacology for investigating interactions between compounds originating from plants and genes linked to different diseases and illnesses. Researchers can identify linkages between phytochemicals and pathways that traditional biochemistry would overlook [13].

AI models trained on Open Targets data, for instance, may determine that a flavonoid molecule controls oxidative stress pathways associated with neurodegenerative diseases. This ability transforms ethnobotanical leads into molecular hypotheses that may be used in modern pharmaceutical pipelines.

10.4.2 Machine Learning Ledger Orchestration for Drug Discovery (MELLODDY): Federated Learning for Private Information

A major advancement in AI cooperation while maintaining privacy is the MELLODDY project, which applies federated machine learning to drug discovery across multiple pharmaceutical companies. The European Union finances the MELLODDY Consortium that brings together 10 multinational pharmaceutical firms like *Janssen*, *AstraZeneca,* and *Novartis* and several AI startups to train shared models on proprietary data without exchanging the data.

MELLODDY uses federated learning, which involves teaching decentralized algorithms locally at each institution while exchanging only model parameters rather than raw data. This increases the resilience of the model while preserving confidentiality (Fig. 10.2).

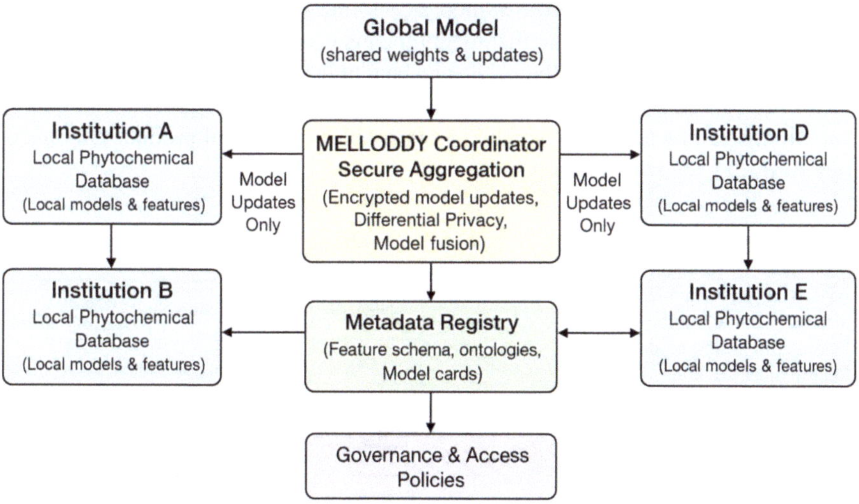

Fig. 10.2 A federated learning system for sharing phytochemical data between universities is part of the MELLODDY platform

Federated learning may connect **regional herbal libraries in phytopharmacology,** enabling AI to learn from diverse ethno-medical datasets across continents while preserving traditional knowledge sovereignty [8]. Global learning would be made possible by such a system, which would preserve ethical purity and local data ownership.

10.4.3 Natural-Product Database Integration

In parallel, the data architecture for AI applications in phytopharmacology is provided by carefully selected natural-product databases. There are two significant examples:

- Approximately 35,000 compounds with annotated bioactivities and organism sources are available in the **Natural Product Activity and Species Source Database (NPASS)** [14].
- **SuperNatural II** has more than 325,000 phytochemicals intended for similarity searching and virtual screening [15].

AI programs that have been trained on these archives have the potential to identify structural counterparts, forecast ADME profiles, and even calculate toxicity [16]. An **interconnected ecosystem of phytochemical intelligence** is created when these databases are integrated with platforms like MELLODDY or Open Targets (Fig. 10.3).

10.4.4 Open Collaboration's Importance for Phytopharmacology

Natural product research becomes more accessible when open-source data and AI analytics are combined. Many therapeutic plants come from low-resource regions, where researchers may contribute data and benefit from joint discoveries. The whole idea of open collaboration in this area really helps with transparency and making sure things can be repeated, which is a big problem in natural product pharmacology stuff, like the reference says.

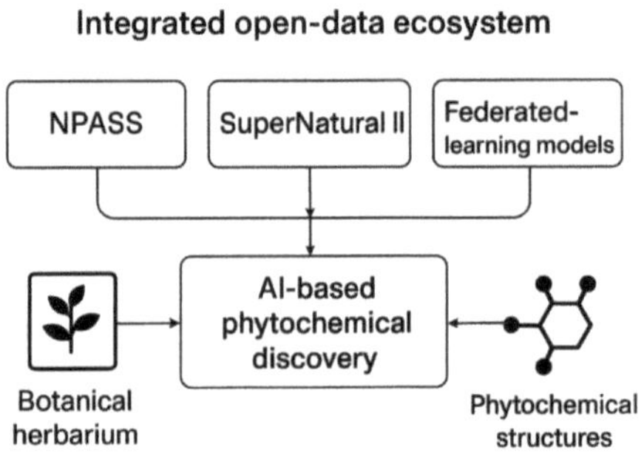

Fig. 10.3 An integrated open-data environment for AI-driven phytochemical discovery that combines federated learning models, SuperNatural II, and NPASS

Pharma companies are kind of mixing things up with a strategy that pulls in data everyone can access, includes ethics from around the world, and uses AI to make it more efficient, blending their own private research with open science setups.

10.5 Collaborative AI-Phytopharmacology Conceptual Framework

The success of AI-assisted drug discovery in the area of phytopharmacology not only requires technical knowledge but also the formulation of collaborative frameworks for the integration of infrastructure, information, and knowledge of various players involved. The conceptual model, as discussed, provides a comprehensive framework for understanding the above requirements related to the interaction of various players involved concerning the development of natural-product drugs.

10.5.1 The Collaborative Ecosystem's Elements

There are five related pillars that provide the AI-phytopharmacology partnership framework:

1. The Data Ecosystem Phytochemical libraries, omics databases, ethnobotanical records, and clinical trial registers are some of the structured and interoperable datasets forming the foundation [14]. The quality and accessibility of this data directly affect the precision of AI models being used by these applications.
2. The layer of AI infrastructure, such as Machine Learning (ML), Deep Learning (DL), and generative AI approaches to bioactivity mapping, compound discovery, and target prediction [1]. This infrastructure is a federated and cloud-based infrastructure, such as MELLODDY, as examples that allow for the multi-organizational training of models without compromising data privacy.
3. Prediction of experiment validation for a chemical experiment by AI. The conjunction of in vitro modeling, in silico modeling, or other methods is necessary for acceptability in regulatory frameworks [5].
4. Partnership Governance: Data sharing treaties, intellectual property concerns, and ethics play a critical role in partnership success. Open science agreements and equitable sharing of benefits options are some of the ways equity is valued in this aspect of partnership [8].
5. Regulatory and Translational Pathways: There must be regulatory interaction in order to analyze the algorithmic decision of potential in phytochemicals developed by AI in making drugs out of those phytochemicals. There also needs standard regulatory coordination between AI companies and the FDA/EMA.

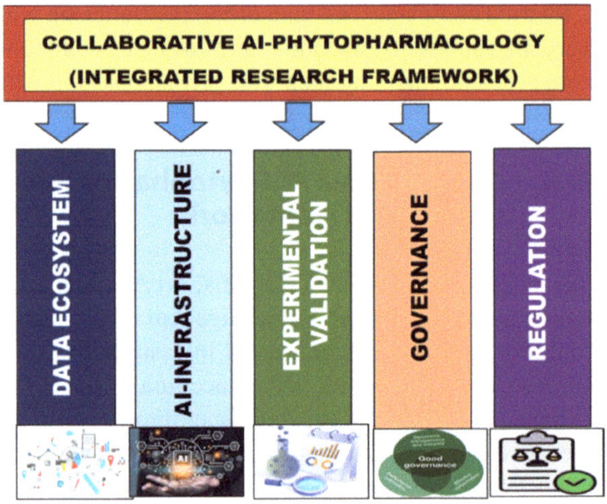

Fig. 10.4 Pillars: Data ecosystem, AI infrastructure, experimental validation, governance, regulation, central hub showing "AI Enabled Phytopharmacology" and arrows showing integration

Together, these pillars form a socio-technical ecosystem that enables AI-driven phytopharmacology to move from isolated experiments to coordinated, multi-institutional innovation (Fig. 10.4).

10.5.2 The Innovation and Knowledge Flow

With pharmaceutical companies providing clinical experience, drug development experience, and understanding of regulations, this model enables a bidirectional flow of knowledge.

- AI companies bring expertise in data analysis and computation.
- Public databases, citizens, and contributors enhance transparency and global coverage of information, while academic/research organizations provide mechanistic knowledge, curate ethnobotanical knowledge, and test confirmatory theories.

With the aim of improving AI models through validation, and thus increasing the efficacy of biological research monitored by AI, a loop is created for co-evolutionary learning.

This leads to the progression of partnership from the transactional type to co-creation, with all stakeholders included in the development, testing, and validation of hypotheses [17].

10.5.3 AI Integration Throughout the Phytopharmacology Process

Every stage of the phytopharmacological process is impacted by AI.

1. Compound Discovery: Using structure–activity correlations (SARs), machine learning identifies bioactive phytochemicals.
2. Target Identification: AI uses chemical characteristics and omics data to predict molecular targets.
3. Deep learning and network pharmacology show complex pathways and processes.
4. Lead optimization: Generative models create analogues with enhanced pharmacological characteristics.
5. Predictive modelling helps with toxicity evaluation and patient categorization in both preclinical and clinical settings.

Each level benefits from collaborative data-sharing systems that allow for cross-validation and iterative improvement [3]

10.5.4 Innovation Outcomes and Cross-Sector Synergy

Pharma, AI, and academia working together yield useful outcomes:

- Candidate screening time may be reduced by up to 60% with AI-driven prioritization (Zheng and associates 2023).
- The range of phytochemicals found worldwide increases therapeutic choices.
- Machine learning shortens the time from bench to clinical trials and speeds up translational research for plant-derived treatments.
- Equitable innovation is encouraged by open collaboration, which guarantees that source countries and indigenous populations enjoy the rewards of discovery.

The Sustainable Development Goals (SDGs) of the United Nations, namely, Goal 3 (Good Health and Well-Being) and Goal 9 (Industry, Innovation, and Infrastructure), which promote sustainable innovation through shared intelligence, are in line with this paradigm.

10.6 Challenges/Difficulties, Ethics, and Future Prospects

Despite the potential, collaborative AI-phytopharmacology has significant challenges in the areas of technology, ethics, and socioeconomics. The field's longevity and global acceptability will depend on how these issues are resolved.

10.6.1 Technical and Data Challenges

Data Fragmentation and Quality: Incompatible databases and publications are still scattering phytochemical information. The performance of AI is hindered by lack of standardization, repetition, as well as unavailability of metadata [15]. Creating standards for FAIR (Findable, Accessible, Interoperable, and Reusable) data is crucial.

Algorithmic Bias and Model Generalization: Because of the increased structural richness in phytochemicals, the AI models that were trained on data from synthetic compounds could not work correctly on phytochemical data [1]. Hybrid algorithms used for both natural products and synthetic compounds in phytochemical analysis can eliminate this problem.

Inequities in Computing Infrastructure Access: The high-performance computing environments required for more advanced AI workflows are also not available in developing countries, where access would be needed due in part to the variety of medicinal plant species, for example. Open-source computing funding could help address.

10.6.2 Ethical and Legal Considerations

Benefit-Sharing and Intellectual Property: The issue of ownership over chemicals discovered by AI that originate from traditional or indigenous knowledge is an important area to consider. Selling genetic or biochemical resources dictates that there is equitable sharing of benefits in accordance with international regimes such as the Nagoya Protocol in 2014. Collaborations involving AI must observe these guidelines to avoid biopiracy in the digital world [8].

Data Privacy and Consent: Even in federated learning, the need to share the data may not be required; however, in ethnobotanical knowledge, issues of metadata and consent are also present [18]. The need for anonymity and open consent exists.

Algorithmic Accountability: It is extremely important to have accountability when it comes to errors or deceptive predictions produced by algorithms, particularly as the influence of AI is seen in the design of drugs as well. Standards for "explainable AI" are already being drafted for regulatory bodies to enable traceability for the path of decisions.

10.6.3 Global Equity and Socioeconomic Issues

There exists an imbalance in the distribution of infrastructure and AI knowledge. Activities like algorithm development and data management are dominated by developed countries, whereas resources and indigenous knowledge tend to be in regions of the world that are hubs of biodiversity. There needs to be an element of capacity building in this partnership, such as developing knowledge within the region, building regional data centers, and revenue streams to source regions to enable sustained ethics [2].

Further, multi-lingual data archives and open access publications could increase engagement with AI-phytopharmacology. International initiatives such as Global Open Science Cloud can address these gaps by emphasizing computational equality.

10.6.4 Prospects/Future Possibilities

AI, quantum computing, and multi-omics integration for the field of phytopharmacology are likely to come together in the next decade itself. AI technology is set to improve the power of quantum computing for accurately modeling the chemical interactions of complex molecules for medical applications. The AI system that can evolve by itself, generate theories, design experiments, and validate these experiments in robotic labs is another area of excitement. These tools hold huge potential in bringing about a paradigm shift in developing drugs derived from natural sources, in combination with phytochemical diversity. In conclusion, addressing these issues is important in ensuring that AI-assisted phytopharmacology remains equitable and scientifically rigorous across the world.

References

1. Bender, A., & Cortés-Ciriano, I. (2021). Artificial intelligence in drug discovery: What is realistic, what are illusions? Part 1: Ways to make an impact, and why we are not there yet. *Drug Discovery Today, 26*(2), 511–524.
2. Newman, D. J., & Cragg, G. M. (2020). Natural products as sources of new drugs over the nearly four decades from 01/1981 to 09/2019. *Journal of Natural Products, 83*(3), 770–803.
3. Goh, S., et al. (2025). Challenges in implementing artificial intelligence in breast cancer screening programs: Systematic review and framework for safe adoption. *Journal of Medical Internet Research, 27*, e62941.
4. Wang, C., et al. (2021). COVID-19 in early 2021: Current status and looking forward. *Signal Transduction and Targeted Therapy, 6*(1), 114.
5. Chowdhury, S. R., et al. (2025). Deciphering the antidepressant potential of Piper betle L. essential oil and its inclusion complex: Network pharmacology and experimental insights. *Molecular Neurobiology*, 1–23.
6. Zeng, B., et al. (2022). Andrographolide: A review of its pharmacology, pharmacokinetics, toxicity and clinical trials and pharmaceutical researches. *Phytotherapy Research, 36*(1), 336–364.
7. Bayer, J., & Högger, P. (2024). Review of the pharmacokinetics of French maritime pine bark extract (Pycnogenol®) in humans. *Frontiers in Nutrition, 11*, 1389422.
8. Jiang, W., et al. (2025). Phytochemistry, pharmacology and bibliometrics insight of the genus Callicarpa L.: An extensive review encompassing the years from approximately 2010 to 2024. *Phytochemistry Reviews*, 1–81.
9. Zheng, L., et al. (2023). The effectiveness of artificial intelligence on learning achievement and learning perception: A meta-analysis. *Interactive Learning Environments, 31*(9), 5650–5664.

10. Philippidis, A. (2024). Trans-Atlantic Triumph: AI drug pioneers recursion, exscientia combine: Company vows first-and best-in-class drug discovery based on artificial intelligence after joining forces to create a pipeline of 30+ programs. *GEN Edge, 6*(1), 950–956.
11. Philippidis, A. (2025). As pipeline advances, recursion expands AI focus to clinical trials: Chief R&D and commercial officer Najat Khan, PhD, tells GEN Edge the company aims to improve design of studies, accelerate enrollment, and enhance evidence generation. *GEN Edge, 7*(1), 63–69.
12. Pun, F. W., Ozerov, I. V., & Zhavoronkov, A. (2023). AI-powered therapeutic target discovery. *Trends in Pharmacological Sciences, 44*(9), 561–572.
13. Yu, C., et al. (2022). Pharmacological effects of polyphenol phytochemicals on the intestinal inflammation via targeting TLR4/NF-κB signaling pathway. *International Journal of Molecular Sciences, 23*(13), 6939.
14. Zeng, X., et al. (2018). NPASS: Natural product activity and species source database for natural product research, discovery and tool development. *Nucleic Acids Research, 46*(D1), D1217–D1222.
15. Banerjee, M., et al. (2021). The impact of artificial intelligence on clinical education: Perceptions of postgraduate trainee doctors in London (UK) and recommendations for trainers. *BMC Medical Education, 21*(1), 429.
16. Xiong, G., et al. (2021). ADMETlab 2.0: An integrated online platform for accurate and comprehensive predictions of ADMET properties. *Nucleic Acids Research, 49*(W1), W5–W14.
17. Sakinc, M. E., & Gleadle, P. (2025). A stakeholder perspective of value creation versus value destruction under financialization: The cases of Bayer and Merck compared during 2000–2022. *Competition and Change*, 10245294251324782.
18. Oldenhof, M., et al. (2023). Industry-scale orchestrated federated learning for drug discovery. In *Proceedings of the AAAI conference on artificial intelligence*.

Part V

Challenges of Artificial Intelligence in Phytochemistry and Drug Discovery

Ethical Concerns in Leveraging AI for Phytochemistry

11

Naima Shahzadi, Raheem Shahzad, Ayesha Batool, Shifa Shaffique, Sobia Ikram, and Adeeb Shehzad

Abstract

Artificial intelligence (AI) has become a paradigm that has changed the manner in which plant-derived products are detected, characterized, and utilized in botanical sciences, phytochemistry, and pharmaceutical research. This chapter summarizes the whole subject of AI-based innovations including machine learning and deep learning along with advanced computational chemistry and multi-omics integration, which are speeding the discovery of phytochemicals and enhancing agricultural and drug discovery processes. The integration of AI with quantum mechanical methods including the Density Functional Theory (DFT) and molecular dynamics has significantly improved the accuracy of molecular modeling and structural clarification and prediction of bioactive metabolic pathways. Moreover, the Internet of Things (IoT) applications powered by AI have transformed digital agriculture enabling it to make data-driven decisions, opti-

N. Shahzadi
Department of Botany, Government College Women University Sialkot, Sialkot, Pakistan

R. Shahzad (✉)
Department of Horticulture, The University of Haripur, Haripur, Pakistan
e-mail: raheem.shahzad@uoh.edu.pk

A. Batool
School of Ecology and Nature Conservation, Beijing Forestry University, Beijing, China

S. Shaffique
Department of Applied Biosciences, Kyungpook National University, Daegu, Republic of Korea

S. Ikram
South Australian Research and Development Institute, Adelaide, SA, Australia

A. Shehzad
Biodiversity Research Centre, Dhofar University, Salalah, Oman

Y. Bibi et al. (eds.), *Artificial Intelligence in Phytochemistry*,
https://doi.org/10.1007/978-981-92-0875-3_11

mize resources, and enhance crop productivity. Even though these developments have been made, significant issues such as the bias in data, lack of consistency in datasets, lack of model interpretability, potential ethical issues, data privacy, and workforce training still pose significant challenges to the full-scale implementation of AI in phytochemical and botanical studies. New ethical principles of the worldwide organizations and the necessity of transparency, accountability, and interdisciplinary cooperation are also mentioned in this chapter. In the future, AI-driven phytochemistry can be improved with the creation of quality worldwide databases, explicable AI frameworks, and hybrid frameworks of computation that incorporate plant biology, quantum mechanics, and machine learning. Through responsible innovation and scientific rigor, AI can unlock the huge chemical library of plants, speed up the discovery of natural products-based drugs, and enable sustainable development in the decades to come.

Keywords

Artificial intelligence · Phytochemistry · Machine learning · Deep learning · Multi-omics integration · Computational phytochemistry · Ethical AI · Algorithmic bias · Drug discovery

11.1 Introduction

Artificial intelligence (AI) is a growing field of application in various fields of science over the past few years, with pharmaceutical research being no exception, as it is now practiced in a vast variety of activities. This movement in direction to AI-driven approaches is heavily driven by the fact that the digitalization of data has increased significantly and created huge datasets that are best appropriated to AI-driven algorithms that are most efficient with large data sets as inputs. Despite the lack of a worldwide recognized definition of AI, it can be widely perceived as the process of simulation of human intelligence that is performed by the means of machines that can think, learn, and behave in such a manner that can resemble the process of human cognition. Consequently, AI systems are able to process and comprehend complex information, draw relevant conclusions, make informed decisions, process environmental inputs, and become progressively more sophisticated through learning-functions, features that have transformed AI into an inevitable instrument in contemporary areas of research [21].

Phytochemicals have received a lot of interest over the years because of their impressive therapeutic potential and great commercial interest. Besides their medicinal use, these natural metabolites are needed to protect plants, regulate intercellular interactions, or to adjust plants to their environment. They also play a significant role in ensuring physiological homeostasis, thus, normal growth and development of plants, which finally affects the crop yield [1]. It is approximated that the number of plant metabolites is almost 1 million which is the significant degree of chemical diversification still under-explored. This large pool of phytochemicals has enormous potential in terms of innovation in the agricultural field, medicine, and other industries [2].

The purpose of this chapter is to critically discuss the ethical issues that come about when artificial intelligence (AI) is introduced into the field of phytochemistry and botanical studies. It is dedicated to the effect of machine learning, deep learning, computational modeling, etc., on data interpretation, phytochemical profiling, and natural product discovery and addresses risks associated with creating AI-based tools: algorithmic bias, data quality, model transparency, and predictive fairness. Issues related to the privacy of data, reproducibility of the research, and responsible use of AI-generated predictions of phytochemicals are also assessed in this chapter. Additionally, it discusses the consequences of adopting AI in terms of training of the workforce, reliance on artificial intelligence, and the possibility of misusing AI-driven knowledge. Finally, this chapter seeks to provide ethical standards, best practice, and future prospects of applying AI safely, responsibly, and trustfully in plant sciences and phytochemistry.

11.1.1 Overview of Machine Learning and Deep Learning Approaches

Machine learning (ML) is one of the fundamental fields of artificial intelligence that implements algorithms that can execute a particular task, learning it by the data, which can be labeled (supervised ML) or unlabeled (unsupervised ML). Deep learning as an advanced branch of ML allows machines to learn in several hierarchical processing layers, which automatically determine complicated patterns in large data sets. Convolutional Neural Networks (CNNs) and Recurrent Neural Networks (RNNs) are some of the most notable deep learning networks that have delivered outstanding performance in image recognition and prediction using data [3, 4].

11.2 Integration of AI and IoT in Digital Agriculture

Digital agriculture is the combination of the latest technologies that have completely changed the agricultural sphere, as the work at farms is now more accurate, data-oriented, and efficient. This change is enabled by the use of automated technologies, including artificial intelligence (AI), the Internet of Things (IoT), and advanced systems of data collection and processing. The AI-based solutions generate and use large amounts of agricultural data to supply data-driven services and decision support systems (DSS) to manage farms better. Besides, this combination of farm-level data and external information sources, especially weather data and environmental data, leads to a higher level of resource optimization, risk management, and crop productivity [5].

11.3 Advances in Computational Phytochemistry and Quantum Mechanical Approaches

The combination of high-throughput computational tools in the past few years has significantly enhanced phytochemical studies. Computational phytochemistry is a wide field referring to a range of computational, mathematical, and statistical methods aimed at helping to intersect theory with experimental results in phytochemical research. It can be applied to different scales, such as the explanation of biosynthetic pathways and a thorough screening of phytochemicals [6]. It is worth noting that the practice of computational methods does not coincide with the emergence of artificial intelligence in the twenty-first century. Density Functional Theory (DFT) is one of the most significant quantum mechanical theories used to model electronic structures of the molecules and their associated properties, including vibrational frequencies, ionization energies, their atomization energies, and magnetic or electric properties. DFT is based on the fact that the quantum behavior of systems can be calculated based on the density of the particles comprising them [7]. As an illustration, [8] have shown a high level of agreement between the theoretical and computational DFT modeling when determining the electronic and spectroscopic characteristics of Pistagremic acid from *Pistacia integerrima*, which successfully estimated the electron affinities and ionization potentials of the HOMO and LUMO orbitals. Likewise, DFT combined with molecular dynamics allows the elucidation of the structure of phytochemicals in organic solvents in detail complementing the analysis of NMR. The 1H NMR spectrum of 5,4′-dihydroxy-7,5′,3′-trimethoxyisoflavone of *Ouratea ferruginea* was also simulated very successfully using the B3LYP hybrid DFT functional, which is another example of how effective computational methods are in phytochemistry [9].

11.4 Role of Analytical Techniques and Machine Learning in Phytochemical Characterization

Analytical methods offer a potent entry point to the discovery of the latent qualities of plant-based samples such as their molecular structure, molecular mass, isotopic patterns, absorbance characteristics, and vibrational modes. Information collected using these state-of-the-art methods is then analyzed (by hand and computational methods) to produce accurate chemical fingerprints of phytochemicals. The recent growth of machine learning (ML)-based analytical chemistry can be explained by a number of reasons. First, the contemporary analytical tools lead to the huge amounts of data, which produces an urgency to engage advanced computational approaches that can identify valuable phytochemical data. Second, the easier availability of high-performance personal computers has allowed plant chemists to incorporate ML tools in their processes. Third, the evolution of information technologies has enabled the intensive data mining, which helps create computational platforms, particular databases, and ML applications in phytochemical studies [10]. Computer-assisted structure elucidation (CASE) of phytochemicals is shown in Fig. 11.1.

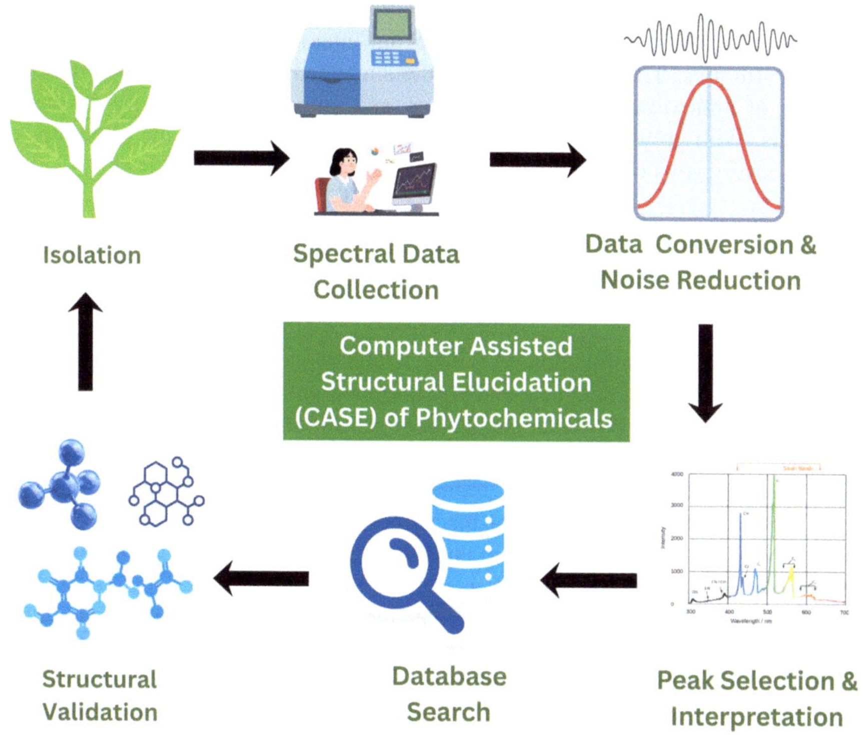

Fig. 11.1 Summary of computer-assisted structure elucidation (CASE) of phytochemicals. The phytochemical collected from spectroscopic instruments will be processed

11.5 Artificial Intelligence in Multi-Omics Integration and Phytochemical Discovery

In order to detect and characterize bioactive compounds within plant systems, artificial intelligence (AI) is more commonly applied to analyze massive omics data, including genomics, transcriptomics, proteomics, and metabolomics. Even though the high-dimensional datasets have rich insights about the molecules, they can be complex, and thus they are not usually amenable to the conventional bioinformatics tools. The algorithms based on AI and machine learning, and, in particular, deep learning models have proven to have an outstanding ability in identifying meaningful patterns and correlations, which, in turn, makes it possible to discover insights into novel metabolites with therapeutic properties. The most important uses of AI are metabolite discovery, structure determination, and mapping of genotype-phenotype interaction to control biosynthetic pathways. Enhanced computational software, including computer-assisted structure elucidation (CASE) and multi-omics platforms, are also used to make more accurate predictions of the secondary production of metabolites and allow scientists to understand the effects of genetic variation on chemical diversity further. In addition to compound discovery, AI has also found extensive applications in virtual screening and drug design, which, among others, predicts the molecular docking interactions and approximates the binding affinities. These methods have demonstrated great potential in in silico analysis of antioxidant, anti-inflammatory, and antimicrobial activity of phytochemicals. With growing capabilities, AI stands to revolutionize the principles of drug discovery at its initial phases based on plant-derived compounds [11].

11.6 Challenges and Limitations of AI in Botanical and Phytochemical Research

Although artificial intelligence (AI) has a massive potential in improving the research in the fields of botanical and phytochemistry, there are a number of crucial challenges that need to be overcome in order to absorb its potential to the maximum. Bias and the quality of the data are also vital problems, since AI models are based on large, accurate, and adequately curated data sets. Inconsistency, incomplete data, and biased sampling can result in the inaccuracy of predictions and models with low quality. Moreover, AI models trained on controlled conditions (laboratory or greenhouse) are likely to have issues when deployed into practice in a real-world field setting, since there exist differences between the environment and the complexity of the biology, which reduce predictive power. The interpretability and explainability of AI models are also another significant problem as, in the pharmacological and clinical fields, it is necessary to have a transparent decision-making process to establish trust and guarantee the ethical use of AI, and explainable AI frameworks are a major research topic (Frontiers in Plant Science, 2023). Moreover, successful cross-disciplinary integration is also a major challenge, because successful use of AI involves smooth cooperation between computational scientists, botanists, phytochemists, agronomists, and ethnobotanists, as well as coordination of heterogeneous and heterogeneous data. These issues are vital to the achievement of the potential of AI-driven innovations in botanical discovery [11].

11.7 Ethical Concerns of AI

According to European Commission, good AI must possess these three qualities:

1. Be legal and respect the law.
2. Be moral and stick to values.
3. It must be sound and true and work as per the technical standards [12].

The companies are deploying artificial intelligence (AI) at a phenomenal rate, but as AI goes mainstream, there have been significant questions on the ethical concerns, which come up out of drug discovery and development process. The most significant ethical issues are being discussed below.

11.7.1 Fairness

Fairness, as one of the fundamental principles of justice, has been a longstanding issue and especially a problem of bias, which has been a lengthy ethical concern over the centuries. The complexity and vagueness of most machine learning (ML) algorithms is a factor that significantly contributes to the problem of fairness because it is challenging to determine whether an AI system is truly fair due to the inability to explain how it arrives at its decision. Among the literature, two most outstanding elements that were associated with fairness-related issues were gender and ethnicity. In the 1970s, a famous example was the case of St George's Hospital Medical School in the United Kingdom, in which algorithmic screening systems had been discovered to favor certain applicants over other applicants based on gender and race. In addition, there are those institutions that deliberately do not gather demographic features like gender or ethnicity due to legal, institutional, or commercial considerations. Although this might seem to uphold the idea of neutrality, the lack of such information might have the unintended consequence of raising the likelihood of indirect discrimination, which will destroy fairness [13].

11.7.2 Algorithmic Bias Has the Potential to Negatively Impact Certain Social Groups

Artificial intelligence involves computers performing the evaluation of the information on the world, presentation, and drawing of conclusions based on this information. Nevertheless, it may be biased adversely affecting social groups, in particular, in the sphere of health, and make erroneous judgments [14]. The parameters, the creator and their end, are already in the society and contribute to inequalities that support bias. The problem of every algorithmic bias is a social issue that unites all the existing differences, socioeconomic status, race, ethnic origin, religion, gender, disability or sexual orientation, and numerous others. The qualitative and quantitative properties of the groups can vary every time the algorithm is used [14]. Nevertheless, considering the problem of algorithmic bias as a mere technical problem, we can only find engineering solutions, and unless the underlying problem is addressed, the disparities of facts and values [15]. As long as our view on our questions is narrow, then that will be embodied in data whether directly or indirectly in the final decisions and final results. Addressing the problem of algorithmic bias requires minimizing it, as there are data and value inequalities in our society, and even the existing laws cannot eradicate them. The less specific an algorithm is used, the more significant the threat of harming social groups because of its delicacy and specialization [15, 16]. Interdisciplinary research groups can streamline algorithm development to maximize performance while reducing bias and ensuring more equitable AI choices.

11.7.3 Data Privacy

With the growing complexity of medical data, and the pharmaceutical and technology industry having need to exchange large amounts of data, data security issues remain an escalating concern. There is a plethora of risks that come with the usage and transfer of sensitive data, and these include indirect information leaks, data poisoning, linkage attacks, reconstructing data using published results, adversarial examples, transferability attacks, and even stealing models. Nevertheless, even with these dangers, a vast majority of patients are ready to provide their personal information to be collected and used to promote the progress of healthcare and research. Nonetheless, the consent can easily add other ethical and practical complexities because patients can be asked to sign general terms and conditions without the full realization of what will happen to their data in future or how their information may be utilized in the second place. Data privacy is thus of paramount importance during AI model training and validation, and several security-enhancing approaches are developed. Also, it is necessary to assess AI systems in clinical settings in real time to highlight the weaknesses of the models and enhance security against abuse or exploitation [17].

11.7.4 Ethical Data Practice in AI Development

In terms of data, ethical AI development necessitates that the data collection, its processing, and further analysis should be performed mindfully, objectively, and diversely. Good, unbiased datasets are the basis of good and trustworthy AI systems, and it is essential to shun away skewed representation practice or perpetuating the biases that exist already in the society. As well, collaboration between the participants of the multidisciplinary team, including ethicists, social scientists, regulatory professionals, and technical developers, holds the key to designing AI systems that would meet the expectations of the wider ethical, social, and legal communities. This collaborative involvement is more responsible in the data governance and makes AI technologies more trustworthy, in general [18].

11.8 Workforce Training Challenges in the Era of AI-Driven Pharmaceuticals

Now that the pharmaceutical industry is quickly becoming digital, it has become a major challenge to retain a workforce that is able to keep up with the changes in technology. The contemporary world has seen firms at all levels of their structure demand that such staff be well trained in artificial intelligence (AI) and how to use it, in order to stay competitive [19]. The rapidly changing processes in the industry, regulatory provisions, and the technological aids require short and regular trainings to facilitate compliance and efficiency. Also, with the further introduction of AI into the medical sphere, there is an increasing demand of specialized skills, specifically in computer science and data analytics [20]. Therefore, interdisciplinary teams that create AI- and machine-learning-based solutions when working on clinical trials, personalized medicine, and drug discovery should have not only basic programming skills but also a basic understanding of biology and chemistry. This cross-functional knowledge plays an important role in filling this gap between computational innovation and biomedical applicability.

11.9 Opportunities and Concerns of AI in the Future of Employment and Medicine

Although it holds enormous possibilities of saving lives and improving business performance, artificial intelligence (AI) still makes people worried about the displacement of careers. Most of the jobs that were done by humans are being automated, and there are concerns that AI will kill the traditional jobs. Nevertheless, it is also necessary to note that AI opens up new employment opportunities since human experience is needed to train, control, and refreeze AI systems. Properly deployed, AI working in cooperation with human workers but not substituting them will computerize the repetitive tasks to have employees devote more time to more meaningful, creative, and high-value work. The pharmaceutical field is one sector where AI has managed to execute activities that are beyond human mental capacity. The development of drugs is, infamously, both very time-consuming and expensive, and the failure rate of drug development is high, particularly in late stages of drug development. These losses can be reduced with the help of AI that will analyze large quantities of medicine and health data, increase the accuracy of the predictions, and decrease failed trials. Additionally, AI can help determine in advance who to include in clinical trials, simulate the behavior of molecules to predict the physiological response, and find genetic biomarkers to justify personalized medicine. These features make AI a revolutionary technology in expediting the process of drug discovery and streamlining the therapeutic process [17].

11.10 Summary

Artificial intelligence has become one of the most disruptive technology agents of the twenty-first century and is changing scientific inquiry in all fields—botany, phytochemistry, agriculture, and pharmaceutical sciences. As it has been shown in this chapter, AI-based instruments have allowed researchers to study intricate biological systems with a degree of accuracy, effectiveness, and specificity that was, until recently, impossible. Since machine learning as the method of predicting metabolic profiles and deep learning as the method of multi-omics integration, AI has already taken its place in the role of speeding up the process of natural product discovery, refining crop improvement plans, and updating the methods of analytical phytochemistry. Specifically, the convergence of AI and powerful computational methods, including Density Functional Theory, molecular dynamics, and computer-assisted structural elucidation, has contributed greatly to deciphering plant metabolic networks, forecasting chemical behavior, and certifying molecular interactions applicable to pharmacology and drug development.

However, there is also a range of scientific, practical, and ethical issues that becomes apparent due to the swift incorporation of AI into botanical and phytochemical sciences. Such problems as data bias, inadequate quality of the datasets, low model interpretability, and the inability of laboratory-trained models to be translated into practice in the field setting may jeopardize the outputs and diminish the trust in AI-based outcomes. Besides, the standardized, curated, and interoperable databases are even more necessary as the multi-omics datasets are becoming increasingly large and complicated. In the absence of strict data stewardship and transparency in methodology, AI systems will become a source of reproducing existing inequity, false biological interpretations, or incorrect predictions.

The ethical issues also make the situation more complicated. Algorithms biasedness, data security, and the ambiguity of patient consent and reuse are the major threats to responsible AI implementation, especially in pharmaceutical research, where human health is at stake. As AI is integrated into the drug discovery process, clinical trial design, and decision-support systems, the necessity to overcome the frameworks that allow ensuring fairness, accountability, transparency, and security appear. It is also vital that the technologically competent workforce must be trained to meet the interdisciplinary requirements of the AI-driven scientific settings. The next generations of researchers will need a solid background in all of biology and chemistry, as well as statistics, computer science, and data analytics to utilize AI to its fullest.

Nonetheless, the possibilities of AI are unparalleled despite these challenges. AI has the capability to discover novel phytochemicals, delineate biosynthetic programs, simulate chemical reactions, and optimize agricultural activity, all of which are the basis of future achievements in the natural product pharmacology and sustainable plant sciences. Artificial intelligence can cut down on drug discovery cycles, experimental workloads, and come up with insights that will bridge the gap between customary systems of knowledge and contemporary biological studies. Responsible implementation of AI does not displace human knowledge but instead supplements it—allowing scientists to devote their creative, critical, and high-value problem-solving skills.

In essence, further AI applications in botanical and phytochemical research will be based on collaborative and ethical, and open innovation. Researchers can make sure that AI will further develop as a transformative but responsible scientific instrument by developing explainable AI frameworks, developing stronger interdisciplinary partnerships, and investing in equitable and high-quality data infrastructures. With the borders between computational intelligence and biological science continuing to be broken, AI is destined to open a new world of discovery, expanding on the chemical richness of plants, deepening the boundaries of natural product-based therapies, and making a significant contribution to the progress of the global health and sustainable development.

11.11 Future Recommendations

The future evolution of the AI-based botanical and phytochemical research depends on the improvement of the quality of data, the model transparency, and interdisciplinary collaboration. To begin with, standardized high-quality phytochemical and multi-omics databases need to be established in order to minimize discrepancy, strengthen the ability of models to be generalized, and facilitate reproducible studies. The combination of curated datasets of different ecosystems, traditional knowledge sources, and worldwide research networks will expand the range of chemicals that can be explored using computational methods even further. Second, there is a need to pay greater attention to the creation of explainable and interpretable AI systems that would help researchers, clinicians, and regulatory authorities to understand how models make decisions, thereby improving trust, safety, and ethical responsibility. Third, literature in the future ought to contemplate hybrid modeling methods, that is, AI with quantum mechanics, mechanistic plant biology, and molecular simulations, in order to enhance predictive accuracy in metabolite identification, pathway mapping, and drug design. Also, the topic of ethical issues related to the privacy of data, algorithmic discrimination, and equitable distribution of benefits with indigenous populations also needs to be included in the baseline of AI development in plant science. Finally, interdisciplinary training programs that cross the lines between the computer science, phytochemistry, genomics, and pharmacology will also be required to develop a properly trained workforce that will be able to apply AI innovations. With such priorities in place, the field can be shifted into more resilient, ethical, and meaningful aspects of artificial intelligence incorporation in the discovery of phytochemicals and the development of plant-based drugs.

References

1. Isah, T. (2019). Stress and defense responses in plant secondary metabolites production. *Biological Research, 52*(1), 39.
2. Saito, K. (2013). Phytochemical genomics—A new trend. *Current Opinion in Plant Biology, 16*(3), 373–380.
3. Arnold, L., et al. (2011). An introduction to deep learning. In *European symposium on artificial neural networks (ESANN)*. ESANN.
4. Pouyanfar, S., et al. (2018). A survey on deep learning: Algorithms, techniques, and applications. *ACM Computing Surveys, 51*(5), 92.
5. Dara, R., et al. (2022). Recommendations for ethical and responsible use of artificial intelligence in digital agriculture. *Frontiers in Artificial Intelligence, 5*, 884192.
6. Sarker, S. D., & Nahar, L. (2024). *Computational phytochemistry*. Elsevier.
7. Trappe, M.-I., & Chisholm, R. A. (2023). A density functional theory for ecology across scales. *Nature Communications, 14*(1), 1089.
8. Ullah, H., et al. (2014). Density functional theory and phytochemical study of Pistagremic acid. *Spectrochimica Acta Part A: Molecular and Biomolecular Spectroscopy, 118*, 210–214.
9. Hernandes, I. S., et al. (2020). Conformational analysis of 5,4′-Dihydroxy-7,5′,3′-trimethoxyisoflavone in solution using 1H NMR: A density functional theory approach. *The Journal of Physical Chemistry A, 124*(25), 5182–5193.

10. Debus, B., et al. (2021). Deep learning in analytical chemistry. *TrAC Trends in Analytical Chemistry, 145*, 116459.
11. Saripalli, H. R., et al. (2025). *Journal of Population Therapeutics & Clinical Pharmacology*. Mikky Publications.
12. Lu, H.-H., et al. (2023). Continuous improvements and future challenges of air pollution control at an advanced semiconductor fab. *Aerosol and Air Quality Research, 23*(5), 230034.
13. Vellido, A. (2019). Societal issues concerning the application of artificial intelligence in medicine. *Kidney Diseases (Basel), 5*(1), 11–17.
14. Panch, T., et al. (2019). The "inconvenient truth" about AI in healthcare. *NPJ Digital Medicine, 2*, 77.
15. Sánchez-Monedero, J., et al. (2020). *What does it mean to 'solve' the problem of discrimination in hiring?* (pp. 458–468). ACM.
16. Mehrabi, N., et al. (2021). A survey on bias and fairness in machine learning. *ACM Computing Surveys, 54*(6), 1–35.
17. Livieri, G., et al. (2024). Ethical issues arising from the use of AI in drug discovery. *Journal of Politics and Ethics in New Technologies and AI, 3*(1), e37093.
18. Wiens, J., et al. (2019). Do no harm: A roadmap for responsible machine learning for health care. *Nature Medicine, 25*(9), 1337–1340.
19. Aydoğdu, S. (2012). *Training's effect and application to improve employee efficiency in pharmaceutical industry*. APMBA.
20. Hazarika, I. (2020). Artificial intelligence: Opportunities and implications for the health workforce. *International Health, 12*(4), 241–245.
21. Lamberti, M. J., Wilkinson, M., Donzanti, B. A., Wohlhieter, G. E., Parikh, S., Wilkins, R. G. & Getz, K. (2019). A study on the application and use of artificial intelligence to support drug development. *Clinical Therapeutics*, 41(8), 1414–1426.

Legal and Regulatory Implications of AI Utilization in Drug Development

12

Muhammad Salman Khalid, Raheem Shahzad,
Luay Rashan, and Adeeb Shehzad

Abstract

Artificial intelligence (AI) and machine learning (ML) models have revolutionized the field of medicine by optimizing the protocols required for screening and development of novel phytochemical compounds with the help of trained machine learning models. These machines, along with deep learning generative modeling, enable the efficient study of structural modifications, extraction optimization, and screening of plants' primary and secondary metabolites. But still these devices must be strictly regulated in context of several ethical and legal concerns, to guarantee their optimum functioning by minimizing the chance of errors. Data protection, data openness, bias, accountability, data privacy, and patenting are the main ethical concerns that are still to be discussed completely in the context of AI use. The main obstacles encountered by AI-assisted devices are the need for larger datasets, including sensitive healthcare data, which ultimately creates concerns that affect doctor-patient trust, and the inability of these tools to justify the procedures, highlighting accountability. Furthermore, regulatory concerns like legalization, ownership, cross-border data transfer, data collection, and optimization protocols are yet to be fully understood, and competent authorities like EMA, FDA, GDPR, HIPAA, HAS, and MHRA are continuously adapting to overcome these issues, because, if unresolved, patient-related data will be under consistent threat. As many of these trained models are already being hacked by

M. S. Khalid
Department of Allied Health Sciences, IQRA University Chak Shahzad Campus, Islamabad, Pakistan

R. Shahzad
Department of Horticulture, University of Haripur, Haripur, Pakistan

L. Rashan · A. Shehzad (✉)
Biodiversity Research Centre, Dhofar University, Salalah, Oman
e-mail: ashehzad@du.edu.om

Y. Bibi et al. (eds.), *Artificial Intelligence in Phytochemistry*,
https://doi.org/10.1007/978-981-92-0875-3_12

the utilization of new cutting-edge technologies, there is a dire need to train these models with robustness to make them against cyberattacks by integrating federated learning and end-to-end encryption. The focus of this chapter is to emphasize the importance of phytochemical compounds that are primarily used in healthcare sector. Moreover, to fully understand the working of AI-assisted devices and models in healthcare sector, we also highlighted that these challenges must be addressed to maintain transparency and fairness in data collection, storage, analysis, and interpretation.

Keywords

Artificial intelligence · Machine learning · Screening · Optimization · Data Privacy Regulatory Compliance

12.1 Introduction

Phytochemistry is a branch of botanical sciences in which researchers study the phytochemicals produced by medicinal plants. The study is mainly assisted by the identification, isolation, screening, analysis, and chemical characterization (structural modalities) of various key, primary and secondary, metabolites like terpenoids, flavonoids, phenolics, and glycosidic residues present in the plant sample. Apart from this, the field is also integrated with ethical AI to understand metabolism, growth, defense, antimicrobial, anti-inflammatory, and anticancer activities. Artificial intelligence is the branch of computer science that is associated with the creation and automation of such machinery that should be able to mimic human intelligence and intellect to carry out tasks like learning, problem solving, and most importantly, decision-making. It does so by analyzing and interpreting large number of datasets (big data) to detect any sort of scientific patterns related to ecosystem, engineering, and various fields in biological sciences to predict the potential outcomes ensuring better accuracy and faster resolving time [1, 2]. In biological sciences, AI has revolutionized the screening process of phytochemicals from botanical species and has been extensively utilized for drug optimization, ensuring optimum dosage for diagnostic and therapeutic purposes instead of traditional drug discovery that was very time-consuming and hectic. These AI models have been trained to work in proximity with techniques like liquid chromatography mass spectrometry (LC-MS), nuclear magnetic resonance spectroscopy (NMR) to enable rapid identification, screening, storage, and modifications of plant-related data. AI does this by relying heavily on intelligent algorithmic machines that mimic human-like intelligence. These machines help in detecting patterns and help with informed decisions by analyzing and interpreting vast amounts of datasets by sorting out millions of active phytochemicals present in the sample [3]. Apart from just screening potential phytochemicals, these AI tools that also do firsthand analysis of clinical trials data, scientific literatures, existing genetic databases like NCBI, PubMed Central to ensure cross-checking and consistency in the overall processes. Analysis of plant-based chemical moieties based on their properties is the first thing that these AI

Fig. 12.1 AI across the screening and identification of phytochemicals in medicinal plants

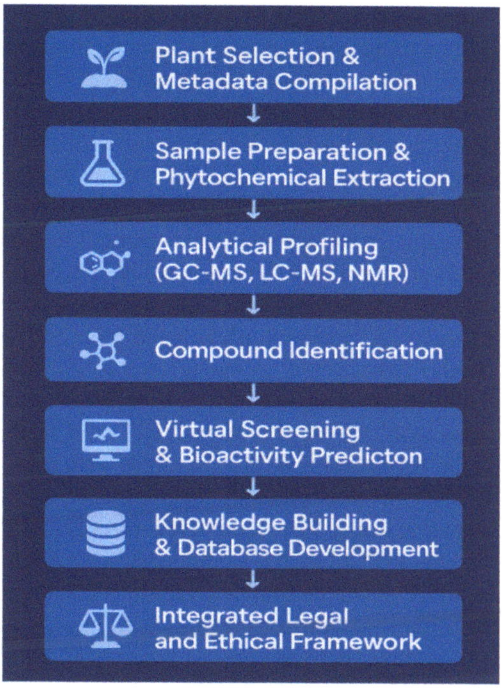

tools study to optimize the chemical modification and design processes (Fig. 12.1). This, in turn, dictates overall efficacy, safety, and reduction in cost and time along with enhanced modifications. In clinical trials, the main issue is to focus on recurring patient data that can be hectic, and it increases the chances of duplication of entries. These tools help in patient identification based on their medical history, making the process a lot of more easier as patients are easier to screen, along with better and optimized dosing regimens ultimately affecting the patients' health positively [4]. Additionally, AI-assisted tools reduce manufacturing costs and improve the production efficiency of plant-based drugs. Beyond the factory, they enable efficient post-market surveillance to monitor drug safety and highlight potential adverse effects.

As the use of AI has increased to a greater extent (extensively utilized in the diseases diagnosis, treatment, phytochemical analysis, and assessment of metabolites, and patient outcomes prediction), it is best to ensure ethical and legal use of these tools that would otherwise raise concerns in the healthcare industry. Judgments, bias, transparency, worker displacement, data protection, privacy and security, liability, and accountability are some of the legal concerns that may arise on the use of these AI tools and machines. These concerns will affect the population negatively by making biased judgments, so these are to be addressed initially to avoid any mishaps in society [5]. One of the biggest legal concerns is the fair use of these tools to ensure unbiased results toward specific populations leading to false diagnosis and inaccurate treatment approaches. To avoid this, scientists and researchers must make sure to train the ML algorithms on large datasets and take necessary measures if something goes wrong. Another highlighted challenge is to make sure the data presented by these models are transparent and explainable, as they can sometimes produce complex results that may not be interpreted by the researchers. Healthcare professionals must make sure to provide enough information and evidence to the patients as they have the right to know what is happening within their body, after they are given the phytochemical-based drugs to ensure transparency. Another debatable concern is the replacement of human judgment with the decisions made by these algorithms because the sole purpose of these algorithms is to be used in enhancing screening and optimization of secondary metabolites [6]. Workers should be trained to ethically utilize these tools and be informed of the potential negative aspects (what could go wrong) as this will restrict them to using AI for improving public health. The healthcare sector must ensure the unbiased and transparent use of these tools, which, if breached, will raise serious concerns. Another major concern is the replacement of human forces with these machines that would be addressed seriously as this will cause the population to riot against the healthcare sectors. The extensive use of these automated AI tools in the pharmaceutical sector will have the chances to displace the workers that were employed by the industry if they were not retrained about the adaptation of the job. Industries should make sure they are mindful of these facts while integrating AI into every domain and should make strategies to minimize these as early as possible [7, 8]. At present, the large amount of collected and analyzed data in the healthcare industry is being done by the ethical use of AI tools, but sometimes the data collector may overlook the patients' rights, privacy, and security of the data. The General Data Protection Regulation (GDPR) in Europe has set some standards for the protection of patient data that has to be practiced in every healthcare industry in Europe to ensure the integrity and inaccessibility of unauthorized access to the big data collected by the companies [9].

Based on the legal and regulatory concerns that might and are being raised on the use of AI in phytochemical analysis, several government bodies have standardized some protocols and legal framework over the extensive use of these tools to address concerns like data quality, security and standardization along with a perfect balance between the opinions of the experts and the results explained by these tools and models. After the establishment of healthcare sectors, these are being operated within a complex network of a lot of national and international regulatory bodies like Food and Drug Administration Botanical Drug Development Guidance (FDA), European Medicines Agency (EU Traditional Herbal Medicinal Products Guidance), and the Nagoya Protocol for accessing biodiversity. These regulatory imperatives were highlighted after the establishment of the International Council for Harmonization of Technical Requirements for Pharmaceuticals for Human Use (ICH) back in 1990 to integrate jurisdiction with the pharmaceutical development. However, the international bodies are also facing challenges over the use of these AI tools in healthcare sector as they defy the traditional paradigms that were set initially for the industries utilizing phytochemicals in drug design and formulation [10,

11]. The model presented by the EMA was structured and entirely focused on risk-tiered approach by providing formalized and clear rules by classifying the use AI in healthcare sector as, "high-risk" (as mentioned in the EU AI Act). The key points of the models include compliance of data with GDPR, mandatory high standards (pre-approval validation), a Multi-Annual Workplan for AI use, and post-market surveillance to ensure the AI is ethically practiced. In comparison to this, the model given by the FDA delivered a dialog-driven, risk-based assessment framework (7 step process), context of use (COU), and flexible model to ensure regulation in pharmaceutical industries by continuous monitoring, maintenance, and modifications of the tools being used. This model was entirely focused on the evaluation of AI based on the context in which it is being used (individual assessment) to ensure bias control and early engagement [12].

Although these AI tools have played the key role in the transformation of phytochemical-practicing industries, there are still some concerns that are still under debate. One of the potential regulatory challenges is the availability of suitable datasets that are free from any sort of errors, as these tools and machines rely heavily on larger datasets (big data) for just training purposes. Initially, these tools must be trained before deploying them for screening the plant samples to ensure accuracy. But most of the time, the data is limited or inaccessible or somewhat incomplete, which can compromise the results [13]. The second thing that raises serious concerns is the ethical considerations that are primarily linked with the fairness of data (bias), as if the data to be trained is only biased to one single species for that specific plant, it will give us unfair results. These issues must be addressed to avoid any typical sort of machine and human error as the future prediction about any specific phenomenon under consideration will be based on the result provided by the machine or the tools that are being utilized. A lot of strategies are currently under consideration to address the fair and unbiased use of AI in the field of biological sciences—chemical medicines (mainly in pharmaceutical and epidemiological studies). The best strategy addressed so far is the augmentation of data that is based on the development and integration of synthetic data with the existing datasets, as this will not only improve the quantity but will also add data diversity, enhancing the datasets to be trained into a favorable algorithm by machine learning [14]. Explainable AI (XAI) is the recently introduced approach that relies on the transparent and fair explanations (judgment on datasets) made by the machine leaning or deep learning algorithm. These approaches are said to surpass the existing approaches that were designed for the traditional methods of experimentation but will never replace human experiences, as these tools only provide predictions on the available data and those predictions (results) are still to be evaluated and validated by human researchers to ensure the fairness of the dataset. By integrating AI tools with human expertise, new approaches will be discovered that will ultimately optimize the metabolite screening and manufacturing processes [15].

12.2 AI Techniques in Phytochemical Analysis

12.2.1 Machine Learning Algorithms for Chemical Profiling

Artificial intelligence has been playing a pivotal role in maintaining the speed, accuracy, and sensitivity of phytochemicals by integrating trained models with ethnobotanical data so that efficient identification, screening, and quantification of metabolites are possible. Traditional screening methods like GC-MS, NMR, and HPLC have been analyzed to produce complex and difficult-to-interpret text that were difficult to interpret by human intellect. With the incorporation of deep and machine learning models with ethnobotanical data, researchers can analyze vast amount of datasets (big data) with ease, ultimately affecting the identification of specific trends in botanical sciences. Chemical fingerprints of medicinal plants are studied easily by introducing supervised learning models, while unsupervised models are currently being used to assess the natural groups in the screened plant metabolites. But still these tools are under strict regulatory compliance because of the repeated unethical practices conducted throughout the globe as can be seen in the USA, where the major concern is the utilization of these phytochemical products. Many of these products are regarded as dietary supplements rather than classified as drugs as per the directions of the US FDA [16]. The regulatory body DSHEA 1994 has standardized the protocols regarding the integration of good manufacturing practices prior to and after the commercialization of these nutraceuticals. To overcome this crucial gap in pre- and post-market surveillance, the AI-assisted devices (integrated with analytical techniques) were introduced to ensure quality and quantity, and to detect consistency and batch-to-batch anomalies that may arise during the entire process. As can be evident from the extensive use of these AI tools, they can also be trained to understand the chemical fingerprints of the phytochemicals to aid in the optimization and modification of chemical structures. So, these AI tools also support regulation by actively working on the safety and purity profile of key metabolites across the globe. While in Europe, Traditional Herbal Medicines Products (THMPD) is the regulatory body that is overseeing all protocols regarding the use of ethnobotanical products, ensuring good manufacturing practices. European Medicines Agency (EMA) has also been working in proximity with the HMPC and has also developed and standardized monographs for phytochemicals [17, 18].

12.2.2 Deep Learning for Structural Confirmation

These machine learning modalities are playing a key role in active screening, prediction, accuracy, and identification of medicinal metabolites from a sample of plants, regardless of the nature of the sample. Predictive modeling and chemometric analysis have been linked with techniques like random forests (used in the prediction and optimization of phytochemical variants based on anomalies), support vector machines (SVMs), artificial neural networks (ANNs), and principal component analysis (PCA visualizes chemical structures, cluster identification, and outlier detections) to ensure smoothness in phytochemical screening of various metabolites with medicinal effects. ANNs are studied to show a human-like learning pattern and by integrating them with phytochemistry, these networks are helping researchers in comprehension and handling of nonlinear relationships between independent and dependent variables. These ANN models are involved in prediction of compounds (based on size of leaf, nutrients, and varying climatic conditions), optimization (extraction with the help of pH, temperature, and duration of the process), evaluation (improving quality and maturity of plant samples) along with detection (screening of compounds from authentic and mixed plant samples) [19]. Support vector machines (SVMs) are also being used in the evaluation of qualitative aspects of medicinal plants by introducing a robust classification system. Identification of species, detection of structural modalities, classification based on demographics, and grading of chemical markers are the core areas that are considered under the umbrella of SVMs [20]. The models themselves are primarily concerned with the detection, screening, and quantification of alkaloids, terpenoids, flavonoids, and phenolic compounds along with their derivatives. The main advantages of these techniques are the bypass of destructive screening techniques employed by the traditional approaches, positively regulating the overall mechanism of phytochemical analysis and the reduction in time, resources, and errors in laboratory-based research. AI-assisted trained models do provide the researchers with rapid identification, authentication, and optimization of herbal plant species used in the medicinal sector. This, in turn, leads to enhanced quality plant-based pharmaceuticals, ensuring safety, compliance with protocols (set by regulatory bodies), product consistency, and adaptability toward the ever-changing environment.

12.2.3 Generative Models for Novel Metabolites

In addition to the aforementioned context, these deep and machine learning models are extensively utilizing automation workflows where their sole purpose is to detect the metabolites in specific plant samples (based on their metabolic and chemical profiles). Apart from the detection of plants by PCA and ANN, computational neural networks (CNN) are also being utilized in the detection of pests and various types of stresses that a plant can encounter over the course of its growth. These AI-assisted modalities, specifically UV-Vis spectroscopy and LC-MS, are being utilized effectively in the plant-based products as they can identify the phytochemical present in a specific sample by making and comparing the spectral lines with the existing databases (recognition of spectral peaks, reduction in noise, and metabolite quantification). These AI models facilitate species classification and the detection of plant stress—including drought, temperature, and pollution—as well as diseases and nutritional deficiencies. By incorporating these models with microfluidics and automated robotics, high-throughput screening is achieved, ensuring the efficient extraction and analysis of multiple plant samples simultaneously. Furthermore, real-time monitoring with the real-world botanical data, virtual screening of metabolites using AI tools, and real-time screening and identification are also some factors on which various countries are actively working to further strengthen the AI in phytochemical analysis [21, 22].

12.3 Legal Considerations in AI-Driven Drug Development

12.3.1 Data Governance and Ownership

12.3.1.1 Data Ownership

In medicine, big data plays the most important role in terms of integrity, authenticity, and reproducibility, and, for that, effective data management techniques are to be utilized to ensure traceability and accessibility of protocols, datasets, and the metadata associated with these. Data organization, storage, sharing, reproducibility, and documentation are the legalized practices that are being used as per the direction of the governing bodies (Table 12.1). It is because low-throughput experimentation in biological sciences is subjected to variations, as has been previously seen in many studies, and this makes this data at risk of breach, along with a chance of bias and certain outliers like errors. AI relies on vast phytochemical datasets, yet ownership of plant-related data remains legally ambiguous. Questions arise about whether indigenous communities, botanical gardens (from where the plants were taken), research labs, or AI developers hold the rights to use such data. Scientific integrity and reproducibility are the two main factors that are to be seen while considering the effective data management techniques, as integrity demands that the underlying experimentation is conducted with honesty and transparency and the protocols that are being used are legally standardized [26, 27]. Reproducibility, on the other hand, relies solely on data management, as the data collected during the research could be replicated and validated to ensure its documentation and organization. Data management is critical for long-term stability and accessibility, and for some confidential datasets, the value of that data is beyond the initial study. The data in biomedical research is crucial for predictive modeling, meta-analysis, repurposing, collaborations, and training of machine and deep learning algorithms, ultimately affecting the efficacy of phytochemical screening and manufacturing along with improved personalized medicine.

12.3.1.2 Data Licensing

Data licensing is also a crucial step for maintaining the integrity of datasets as these datasets are to be accessed worldwide to ensure their reproducibility. Although most of these datasets are open-access, they still need to be protected. There is a need for proper licensing, as these aggregated databases are to be used in reproducibility, but still a lack of licensing and rigorous data reproducibility is seen worldwide. There are some issues like nonstandard licenses, missing licenses, fake licenses, and certain licensing provisions that are being discussed. And to avoid these, a compact was made by the name of Reusable Data Project (RDP), a five-star scoring system that utilizes 56 scientific data resources. The main theme was that the data should be

Table 12.1 Key legal challenges in AI-driven phytochemistry

SR no	Legal area	Key issues	Implications	References
1	Data ownership and accessibility	Lack of identification of the potential entity to whom the data must be propertied, whether it may be the person, hospital, or the company manufacturing that AI-assisted model	There is a need to define the usage, legislative licensing, and IP rights distribution in the context of data licensing agreements (DLAs)	[23]
2	Privacy	Re-identification, GDPR/HIPAA constraints, data localization (laws restrict the data to be used across border)	Consent and de-identification, DUAs, and privacy to protect big data were the major hurdles. Increased infrastructure was observed that increased the overall cost	[9]
3	Intellectual property	Whether AI tools or human beings will be the inventor and the freedom to operate (FTO)	Patent filing strategy was introduced in which the focus was shifted to targets identified by AI along with introduction trade protection Computational FTO tools were integrated to overcome risk of regulatory penalties	[6]
4	Liability and accountability	Accountability for errors like black box (difficulty in tracing the error in trained model) The need to understand unethical AI use by the researcher or scientist (false recommendation)	Innovators face a liability risk if the model is defective or gives false results, needs to necessitate AI governance Introduction of XAI in delivering fair and unbiased judgments in favor of the dataset being utilized. Need for utilizing fair methods	[7, 8]
5	Transparency	Black-box models create a loop where the source of error is difficult to trace because of the lack of information of working of the specific deep learning model	Hinders regulatory approval and trust and to overcome this, the concept of sandbox was introduced	[8]

(continued)

Table 12.1 (continued)

SR no	Legal area	Key issues	Implications	References
6	Regulatory compliance	Model validation (stability of model against slight data variations while utilizing it in pilot studies) Robustness (chances of errors should be minimum) Rise of the model bias that will lead to unfair judgment against the prescribed medicine due to unseen data	GMLP protocols should be practiced for the day-to-day monitoring of the models along with the TPLC approach Biases should be mitigated by following the criterion set by the GDPR, MHRA, and the EU AI Act to ensure proper documentation	[7, 24, 25]

clearly stated, comprehensive, non-negotiable, accessible, and reusable, with clear guidelines on who may have access to these datasets. Furthermore, six different types of licenses are currently employed to maintain data integrity and confidentiality. These licenses are categorized based on the reusability of the data and includes permissive, copyleft, restrictive, copyright, and private pool; their individual uses are scored based on the evaluation criterion RDP [28]. Permissive licensing allows the users to reuse, redistribute, or transform the collected and stored data, while copyleft, on the other hand, requires the addition of new contributors if the data was reduced or modified under the same license category as initially given to the dataset or algorithm itself. Restrictive types of licensing enable the users to get data (from plant databases, various genomic datasets, or metabolomic libraries) after accepting various terms and conditions for ethical use thus creating a robust protocol (no derivative) to prevent data integration. Furthermore, private-pool type of licensing enables the users to add data while limiting the pool to joined users as well. Lastly, copyright creates legal ambiguity by positively asserting the exclusive rights of the innovators [29].

12.3.1.3 Data Localization Laws

The sensitive nature of the phytochemical dataset is the most critical point while addressing the use of AI in the pharmaceutical sector, as confidentiality is to be maintained properly. To overcome this issue, data localization laws were introduced in the pharmaceutical sector. These laws are directed to overcome the aforementioned legal, ethical, and regulatory concerns that might arise from data handling with the help of machines and deep learning algorithms. The integration of GDPR with the widespread data extension is serving as the major law that governs all other 145 laws that were reported back in 2021. This integration is fueled by the mandates of localization and sovereignty of data. The GDPR is responsible for data protection through the framework further endorsed by the FDA and RDP, along with rules and regulations of cross-border transfer of the data to be used and analyzed by the scientific community [30]. The localization of these laws across national borders requires this strategic data to be physically stored, kept, and processed to properly utilize the data in international demographics (epidemiological studies). The above-mentioned mandates are further integrated with laws like economic protection, national security, and domestic access to be completely molded into some rigorous modalities that can strengthen the data protection and privacy. For some international technology and pharmaceutical companies that rely solely on cloud computing, the free flow of the big data along with the ethical use of AI in comprehending the big data—these mandates and regulatory authorities enhance the authenticity while minimizing the overall costly measures (data mining, data, mirroring or any independent data processing company). All these measures help strengthen the phytochemical screening, testing, evaluation, and optimization of secondary metabolites that utilize the AI and machine learning tools [24].

12.4 Privacy and Data Protection

12.4.1 Reidentification Risks

The primary risk factors for privacy and data protection are the data breach (forcefully attacking the cloud servers for obtaining confidential data) and unauthorized access that put the patient data at risk of being exposed globally. The extensive use of AI in the pharmaceutical sector is highlighted as a cause of data breaches, as the server can be shut down completely due to the lack of technical skills and trained professionals who know how to integrate and manage AI with the existing ethnobotanical data. As can be observed, ethnobotanical data (can lead to misuse of tribal medicine) is the most valuable asset in the international black market due to a lot of reasons [31]. This data contains various pieces of information like the origin, phytochemical extracts, availability, and potentially harmful and beneficial impacts of the plant that is being considered against the specific disease, and once it is breached, it can cause serious national concerns like identity crisis, targeted attack in food industries, and misuse of drug in pharmaceutical sector. One of the main targets of cyber criminals is the cloud computer, on which all sorts of data is stored, and these AI tools primarily depend on these large datasets for proper functioning, so any technical or human-related negligence puts the data at risk of being used by any external third party [32, 33]. There are a lot of AI tools that are made and trained by third-party vendors (including search engines) like, tools for model training and strengthening, tools for analyzing and interpreting datasets (data analytics), and as these vendors use cloud infrastructure linked with each other, they offer several other layers of vulnerability. To overcome all the aforementioned security threats, additional layers of security are required such as network monitoring, end-to-end encryption, and multi-factor authentication (two-way authentication), which are advised by the GDPR for all the pharmaceutical industries. If these are not properly installed, attackers can infiltrate and exfiltrate sensitive data, and most of the time, the source of breaches is internal (within the facility) due to lack of trained professionals and malicious or careless workers. This data breach is far more dangerous than any sort of financial loss. This will erode patient' trust toward the healthcare industry, incur legal penalties, and in some cases lead to the shutdown of specific services by the legislative and governing bodies. To ensure all this, a robust and strong training system must be considered by the AI-driven health sectors so that there might be no chance of data theft or breach [34].

12.4.2 GDPR and HIPAA Compliance

So far, the two most influential regulatory bodies that are looking into healthcare AI are the GDPR and Health Insurance Portability and Accountability Act (HIPAA). The first one is already being accountable for maintaining the integrity and privacy of phytochemical data in European countries, and the standard for doing this is very high, as this regulatory body grants individual rights in the context of data protection. Informed and rigorous consent in terms of big data processing, and the obligations to access, collect, analyze, correct, and delete data are some of the main protocols and standards set by the GDPR [35]. This regulatory body binds the respective organization in terms of limitation of data generated and organized, purpose limitation, along with protection of data enforced by limitations of design in the context of specific machine and deep learning automated algorithms. HIPAA, on the other hand, relies solely on the protection of personal health information in the United States only by combining rigorous training protocols for the healthcare professionals that are directly involved in data collection and sorting. The main protocols set by this regulatory body are the end-to-end encryption, training of the workers, documentation of important tasks, and limited access controls toward the stored or organized data. The main drawback of HIPAA is its outdated nature of working, as it is unable to cover mobile integrated AI apps and tools (wearable technologies) [12]. Although these regulatory bodies have regional and demographic differences, the main theme revolves around data privacy, consent, rights, ethical and legal considerations, and transparency to protect the data from any sort of breach.

12.4.3 Cross-Border Data Transfer

While talking about the use of AI in the field of phytochemistry, one of the most crucial aspects is the storage and transfer of the medicinal or poisonous plants data across national and international platforms, as these cloud servers (where the data is stored) are constantly under the threat of being breached. If left unsecured, the data fall into wrongs hands and can be tampered for some unethical practices as can been seen by some recent cases across the Asian countries. Cross-border data transfers require adequate safeguards such as standard contractual clauses and encryption. And for that, not only GDPR and HIPAA are working, but a lot more of other regulatory bodies are actively involved in the selection and screening of data that must be transferred across borders. Countries like Brazil (under the regulatory body LGPD), Canada (with PIPEDA), and Australia (working under the privacy act of 1988) are working under their own data protection laws to transfer or receive any sort of healthcare data to their neighboring countries. All these law enforcement agencies are actively working on identifying potential threats and breach firsthand to avoid any sort of data theft from their respective regions. These law enforcement agencies do so by establishing the concept of end-to-end transcription, where data sent by a lot of servers can only be encrypted by a specific key, similar to the lock and key model of enzymes, along with regular testing and model training [12, 36].

12.5 Intellectual Property (IP) Challenges

12.5.1 Inventorship of AI-Generated Biomolecules

Traditionally, any sort of invention or innovation was said to be the product of creativity and intelligence of human beings, and in most of jurisdictions' laws and regulations, the person filing a patent for any specific invention must have to submit proof of whether he/she has done some work on it or not. So, the results were the ownership can only be held by the person who invented, discovered, or created something, along with the firms in which he/she has done the research. The Patent offices in the United States, the United Kingdom, and EU currently do not recognize AI systems as inventors. This raises questions about ownership of molecules designed autonomously by generative models. The extensive integration and usage of the abovementioned AI tools and machines render the researchers to look in the question, "Who is the real inventor of these AI-generated phytochemical structures?" as these structural modalities are to be created using the computed system (computational drug designing—CDD) [37]. At present, the regulatory and jurisdictional bodies do not recognize machines as the inventor rather than the people who have worked on that specific machine to get information regarding the chemical nature of the plant's metabolites. The main problem in patenting these AI tools is that the machines sometimes generate molecules and required things without any human interference (Automation), rendering the enforcing agencies to think of whether the patent should be done for specific AI modules or not [8, 38].

12.5.2 Protection of AI Models

The need for data protection and privacy can be dealt with by following the standards and guidelines set by the regulatory bodies, but the need to protect these AI models from threats and breaches is yet to be fully understood. For that purpose, a protection strategy was initiated to protect the cloud server along with the algorithms of deep and machine learning, and these include patents, copyrights, and trade secrets. Trade secrets are, by far, the strongest, effective, reliable, and readily available form of data (training data, AI algorithms set by the machine learning, weights of the models as well) protection. Restrictions of unwanted access, confidentiality, and end-to-end encryption are the key mediators in trade secrets. The second one is patents that are used to protect the output given by the AI models, but not the models themselves. This patentable information includes screening or identification of new compounds by AI, active pharmaceutical ingredients (API) validated by models, and the optimization of screening methods used in the screening and manufacturing of drug and other metabolites. The last one is copyright permission that generally protects the overall source code of the AI model from any sort of data breaches or phishing attacks, but the main drawback is its inability to protect the data, model, and the mathematical logical operations being carried out [39].

12.5.3 Freedom to Operate (FTO)

De novo design methods (the incorporation of AI in traditional medicines) are gaining popularity for their effectiveness, but they are raising several questions on the intellectual property of human beings in combination with the freedom to operate (FTO) practices. The integration of these models in the computational drug discovery and design is a necessity, as it is robust, high in output, and automated, but still it is yet to be infringed into FTO analyses. Intellectual property (IP) rights are the only boundaries that are not to be crossed whenever new research, development, optimization, or commercialization of a new product is done in international market, which is also a key consideration in the ethical integration of AI tools and algorithms into the production or optimization of new and novel molecules. The challenge arises because these models have the potential to work (as they use deep learning algorithms) far beyond human intellect, discovering and creating new chemical moieties that may not be reported in the existing families (chemical claims of the researchers), creating an international debate on their use. That is the reason companies must make sure that they are integrating sophisticated AI-augmented algorithms into the design given by de novo model of screening drug molecules [40]. The professionals using these tools must make sure that the molecules to be screened or optimized are more likely to be patented or whether there may be any chances of infringement risk associated with those molecules by utilizing approaches like substructure matches, similarity in binding, and expected biological results. An iterative process must be utilized while dealing with FTOs, as these are not a one-time event (the patent landscape is continuously changing daily along with the adaptability of AI modules) [39, 41]. The FTO method should address two of the basic things, guiding the team of researchers away from the harmful chemicals that may form during the drug manufacturing and providing further evidence of the fact that the methods used were fair and unbiased.

12.5.4 Liability and Accountability

Liability and accountability are the two main factors in the context of healthcare sectors as the person using any kind of practice in a public or private healthcare sector must be accountable while taking a risk or making a decision in using any sort of diagnostic or therapeutic regimen. When AI contributes to decisions that lead to clinical failure, safety issues, or financial loss, determining liability is challenging. Responsibility may fall on developers, pharmaceutical companies, clinical researchers, or data providers. As these tools do not have any intellectual property like human beings, there are chances that they may make decisions or give us results that may be harmful for patients and ultimately damage the protocols leading toward patients' compliance. The main concerns that are arising from the ethical use of AI in medicine are, "if these algorithms make a decision (on wrong identification of chemical moieties, misidentification of medicinal plants or in rare cases prediction of toxic compounds as medical ones) that may leads towards the compromise of patients safety, then who will be held responsible?" as no one will be willing to bear the consequences of losing one's job. If anything happens, the blame will either be shifted on the worker using that tool or the company that created that specific algorithm. The complex challenges arise due to the black-box nature (the inability of human beings to interpret the work, decision-making, and complex nature of the AI algorithms) of the model trained by using deep or machine learning algorithms that are responsible for making the decision difficult, as it is nearly impossible to overcome the phenomenon of causation. To overcome all the concerns, the companies should follow the protocols given by the regulatory bodies (the use of explainable AI modules) in the abovementioned sections [42].

12.5.5 Algorithmic Transparency and Explainability

The use of transparency and explainability in the medicinal sector is very critical. The researchers are relying on the use of AI tools instead of using traditional methods in the field of medicines. There is a strict need to use a well-organized, standardized, explainable, and highly transparent approach for the evaluation of interpretations produced by these AI tools. Black-box models can impair regulatory review and undermine trust. Agencies emphasize explainable AI (XAI) in safety-critical tasks [43]. The use of these black boxes is usually a high-risk approach in the field of medicines, as the decisions made by these tools will have to be standardized for the entire community either in a positive or in a negative way. Furthermore, data about what interpretability means in the context of AI modules in healthcare sector is still under consideration, as researchers are still unable to understand the actual meaning of the term along with how to properly use it to evaluate the metabolic data given by these modules. To overcome these issues, several researchers highlighted the meaning of interpretability in their research, and the main definition that all those researchers concluded was, "interpretability is the presentation of data in such for that can be easily understood by the human intellect." After this development, a medicinal taxonomy was also developed for the rigorous evaluation of these AI tools in phytochemistry that focused on the addition of additional metrics. These metrics made it easier for the developers and the users to interpret the findings given by these deep and machine learning tools, as these metrics highlighted the usage of human oversight (one of the ethical frameworks) to solely use these tools for the judgment not replacement. Although many more principles, beyond than those mentioned above, are still in practice, there are many challenges faced in the implementation of these AI modules [44].

12.6 Regulatory Requirements for AI in Phytochemistry

12.6.1 Good Machine Learning Practice (GMLP)

Good machine learning practices are a set of standardized and optimized practices that are used to ensure the data collected and analyzed using AI tools is reliable, high quality, transparent, unbiased, and trustworthy. GMPL are also endorsed by the regulatory bodies like GDPR, FDA, and EMA in their ability to quickly sort out any type of discrepancies in the data under observation. Although the GMPL standards are still in the phase of maturity, they still have rigorous protocols and standards that make it easier for the researchers to comprehend. The core principles of GMPL include data quality and governance (high-quality data), security (unique encryption), privacy (data must comply with the protocols set by the HIPAA and GDPR), validation of AI models being used (robustness), risk management (context-of-use theory), transparency and accountability (proper documentation), explainability (human-interpretable answers), continuous monitoring (to avoid the phenomenon of model drift), and updates when required [45].

12.6.2 Model Validation

Model validation is by far the most complex phase while using AI tools in healthcare sectors, as these models are to be trained first with the help of big data so that they can function best on the data of the patients when needed. Model validation is one of the key aspects of GMLP, as it acts as a bridge between the experiment conducted (collection, analysis, and interpretation) and the approval needed for conducting that experiment in the laboratory. The validation should be done on a strong and rigorous dataset (even when the model encounters some minor variations or noise in the dataset) to ensure the optimal working of the model and to minimize the chances of errors in further studies. In addition to this, the generalizability, the ability of the trained model to work on a specific population or any unseen data that it may encounter, must be checked to ensure that the model will perform well. All these points are essential if the model it to be approved by the regulatory bodies, as there are some recent models that have given catastrophic results when a small amount of unseen, phytochemical structured data (not included in the original dataset) was added into them, putting the environment at risk. The trained model must be accurate, reliable, specific, and highly sensitive toward its intended use (either in identification or in optimization of a drug molecule) [46].

12.6.3 Reproducibility

Reproducibility or repeatability is one of the major principles used in scientific methods in the field of biology and computer science. The main theme is that the results obtained during any observational or experimental studies should be attained repeatedly with high accuracy and reliability. This is usually done in such a way that the data obtained should be fully transparent, ethical, accessible, and supported by methods that were used for obtaining the data. Any research project can be made reproducible if it is articulated, separated, labeled, stored, and documented in a clear and concise manner, regardless of the methods used for obtaining the data. The integration of AI in phytochemistry is currently under debate because most of the published data lacks both transparency and reproducibility, raising a question mark on the ethical qualities of healthcare researchers [47]. Computer code, unscrutinized methodological details, and biased model training are the key elements highlighted by the critics because they have been transforming the AI-generated data into a typical black box, with no evidence on how the data is presented to the community. To address these challenges, the funders, developers and the users must share the open code (utilization of platforms like GitHub), highlighting the working of that specific deep learning algorithm. Moreover, they must develop a secure and theft-free cloud system for accessing patient-related data, further minimizing the chances of cyber-attacks [25].

12.7 Clinical Development

12.7.1 Regulatory Guidance for Trials

Clinical trials are the gold standard of medical industries and as per the protocols given by the FDA, EMA, and GDPR, the AI-assisted clinical trials should be designed in such a way that there are no chances of bias and unethical practices. If the data used in training these machine models is biased, they may give biased results and will result in discrimination or lack of evaluation of the plant-based drug under evaluation. Most of the time, the clinical data includes test results, physician notes for diagnosing the patient, records for hospitalization, and medical prescriptions. The data is highly classified as it relates to the patient's life-and-death decisions, and if lost, the patient can be stigmatized (if the data is related to mental or life-threatening medical condition). The regulatory agencies are actively working on the development of such work plans that will actively guide the overall process, from data selection to optimization and results. A prime example is if the dataset, on which the deep learning algorithm is trained, is taken from only a certain geographical sector—the results generated will be less effective and, in some cases, harmful for other subjects belonging to other geographical regions of the world. Another problem is the extensive integration of these AI tools in phytochemicals-assisted preclinical and clinical studies, and this high volume of AI-driven biomedical research is accepting the assessments criteria given by the national and international competent authorities as they are facing difficulties in determining how these drug induction trials are designed and managed using these AI tools. To overcome this issue, companies should instruct their researchers to train the model on large, diverse datasets that will be free of any bias as the model could be rejected by the competent authorities [48].

12.7.2 Digital Health Technologies

As we are stepping into an era where AI has penetrated almost every aspect of our life, there is a need to have a strict check-and-balance strategy on the ethical and safe use of AI in healthcare sectors. To further strengthen the healthcare sectors, the internet of things was introduced into the medical industry in the shape of wearable devices. These devices mainly include smart watches, glucose monitoring device, fitness trackers, automated drug delivery patches, and automated inhalers that are involved in the continuous monitoring of a patient's medical condition after the treatment regimen, along with the collection and storage of real-time data in cloud software. Monitoring of heart rate, plasma glucose levels, patterns of sleep, and blood pressure—along with the physical activity of heart and muscles—is one of the major applications of these wearable devices. Furthermore, these AI-integrated devices also alert healthcare professionals to the medical conditions of the patients, whether they are experiencing arrhythmic cardiac activities or fainting. Based on the evidence, these AI devices promise a bright future toward the concepts of public health along with personalized medicine (treatment based on the individuals' genetic profile). Still the main pitfall in practicing this approach is that the data to be collected and analyzed is done outside the hospital premises, rendering it vulnerable to any sort of unauthorized access and data breaches as can be observed by the lack of encryption imposed on these wearable devices. As the integration of AI modules increases, there is a growing need to organize robust data screening, storage, and analysis methods by the utilization of end-to-end encryption and best data governance practices, as instructed by the regulatory bodies [49].

12.7.3 Post-Marketing Surveillance (PMS)

It is the process where the administered marketed drugs are analyzed for any sort of adverse drug reactions (ADRs) in the context of their applications in the field of pharmaceutical sciences. As most of these chemically modified biomolecules usually do not pass through the phase 4 clinical checkpoint, these PMS studies are therefore regarded as phase 4 studies. Post-marketing is the only assessment criterion that lets the scientists discover whether there are common or rare chances of ADR against the marketed phytochemicals. Pharmacovigilance (surveillance of drug safety) and pharmacoepidemiology (population-based assessment of drug usage) are the two fields associated with the detection of ADRs in society [50]. The short duration of clinical studies in combination with the limited availability of time is one of the main obstacles in discovering any sort of ADRs in the post-marketing surveillance. Initially, it was hectic to conduct PMS as it was a bit difficult for the researchers to analyze the whole community at a time, and if they did so, there was not enough human judgment to decide whether the medicine has any reactions or not. To overcome this issue, the integration of AI in PMS was done with the help of electronic medical records and special computer systems that were to be used only in healthcare settings. This integration is under the judgment of regulatory bodies, along with the companies and sectors that are using these systems to analyze and interpret any sort of ADRs through rigorous screening selection and take necessary actions [51].

12.7.4 Automated Signal Detection Risks

The main issue encountered in using pharmacovigilance studies is that these cannot provide a clear picture of the low-risk areas, underreporting of ADRs, and difficulty in the quantification of the overall risk associated with the medicines. To overcome this challenge, the US FDA Adverse Event Reporting System was introduced; it must be primarily used in the collection of reports against the reported ADRs, ultimately supporting the PMS, while in Europe, EMA has been responsible for documenting any sort of adverse reactions associated with a certain drug. Lately, the AI modules have also been integrated to further support pharmacovigilance studies. Although the output of integrating these AI modules is very high, still there are concerns that need to be addressed, as the real-world data (RWD) on which the datasets are trained is very vast, spontaneous, and variable, which may cause these tools to analyze the information wrongly if some unseen dataset from any community is introduced. One of the main concerns with this real-world data is its inability to bypass false positive results (it may show relation between the adverse event and the drug that was being used for that event, but in reality there was nothing) ultimately leading toward patients' withdrawal, panic, or, in rare cases, regulatory disapproval. Adversely, this system can also detect a false negative case, and the researchers will be unable to differentiate whether there was any novel integration of the drug with the metabolites of the human body, raising a question mark in the advancement of drug discovery [52].

12.7.5 RWE Integration

In addition to the false positives and false negatives, these automated signal detection systems (ADS) are highly susceptible, due to algorithmic biases and quality issues in big datasets, to unwanted authorization. Furthermore, the EHRs concentrated in only one demographic region will automatically reflect the bias in the data, also raising concerns about their ethical use. And, if the machine or deep learning algorithmic model is trained only on the dataset of a single population, it will give unfair results and its sensitivity toward management of diverse populations will be compromised (breaching the core principles of pharmacovigilance and pharmacoepidemiologic studies). To date, real-world data has been integrated into the surveillance of chronic diseases and multimorbidity conditions, as most of the time patients dealing with one chronic disease are dealing with another chronic disease, which is evident from the data collected lately by many researchers in Italy [52]. This data has already been stored in databases like the NHS administrative database and plays a key role in managing the regulatory approval for new drugs. Various epidemiological concepts like incidence, prevalence, morbidity, and disease morality are to be discussed in context of the RWE as per the direction given by the competent regulatory authorities. In addition to this, the integration of RWE will enable the researchers to accurately identify, screen, and optimize novel drug molecules for fighting against the issues of antimicrobial resistance (AMR). There is a need to implement strict protocols on the ethical and unbiased use of these AI tools in the study of public health, and regulatory bodies like the FDA and GDPR are standardizing their rules and regulations continuously for a better understanding of patient-related data [53].

12.8 Comparative Global Regulatory Landscape

12.8.1 United States (FDA)

The United States has set standards for the operation of the pharmaceutical companies across the national and international borders, as discussed in the above sections, by acting closely with the GDPR for maintaining data integrity and accessibility throughout. De novo pathway is the only mediated review through which any new plant-based drug is to be commercialized in the market, but still, premarket approval and 510 k clearance are currently used because of their easy-to-go nature while de novo design is strict in its protocols. This design is usually employed for the general safety of the device or apparatus that must be approved by combining general and special controls (if needed), providing the user with safety and integrity by class 1 and class 2. The petition that is to be submitted must have all the necessary information related to the device's purpose, description, ethical considerations, possible errors, and the data that supports its proper working. Based on the criterion of effectiveness, risk, and safety of the manufactured device, the FDA can either reject or accept the patent. While PMA mechanism is very strict in its working that is also employed by the food and drug authority in classification of the products and machines that falls in the category 3 (while de novo only deals with class 1 and 2). Class 3 devices are those that can possess a high risk to the human population, if used unethically or improperly. To obtain the premarket approval (PMA), the owner of the manufactured product must submit a comprehensive application to the regulatory bodies, encompassing valid scientific reasoning and justification about the working mechanism of that specific product to make sure the product is people-friendly in its working. Last but not the least, the premarket notification 510 k clearance also deals with all the aforementioned classes of products that usually have nothing to do with the PMA phase. A predicate must be submitted in favor of the regulatory authorities (510 k clearance to US FDA) to establish the safety and efficacy of the device under consideration [54]. The process itself is applicable to only certain products, including repacked and relabeled, and products intended for domestic use along with products that are to be imported and exported from the US market. Recently, the concept of Software as a Medical Device (SaMD) was introduced, consisting of independent software intended to diagnose, treat or manage health conditions related to plant based medicinal products. However, much work must be done by the competent authorities in the United States to maintain cross-checks on the manufacturing and supply of these machines (Table 12.2) [61, 62].

Table 12.2 Summary of regulatory approaches in major regions

SR no	Region	Regulatory framework	Key elements	References
1	USA	FDA AI/ML Action Plan, FDA Botanical Drug Development Guidance, DSHEA (1994)	GMLP, continuous learning, SaMD, innovation centric, credibility and assessment plan	[9, 10]
2	Europe	EU AI Act, CBD, EU Biodiversity Strategy 2030	Risk-based approach, strict privacy, legislative approach, rigorous validation of trained model	[55]
3	UK	MHRA AI assurance, MHRA traditional herbal registration (THR) scheme	Principle-based, agility, transparency, bias mitigation, model drift, fairness and continuous monitoring	[56]
4	China	Cybersecurity law, NMPA, traditional Chinese medicine law	Data localization, algorithm review, technical guidelines, data localization	[57]
5	Japan	MHLW, PMDA, Kampo medicine regulatory framework	Speed and clarity, model validation and continuous improvement, pro-innovation regulatory flexibility	[58]
6	Singapore	HAS	Risk-proportionate approach, clinical safety, and algorithmic fairness	[59]
7	United Kingdom	GDPR, WHO traditional medicinal strategy	Data protection, privacy, fairness, transparency, explainability and accountability, rights to explanation	[24]
8	Europe	EMA, WIPO, Traditional Herbal Medicine Products Directive (2004/24/EC)	Scientific and structured approach, TPLC framework, drift guidance, reflection paper, pre-market validation (PMA), risk-tiered approach	[56]
9	Asia-Pacific	APAC, Nagoya protocol (APS)	Fragmented approach, National Laws of China and Japan, PMDA, rapid review pathways, technical guidance	[60]

12.8.2 European Union

Apart from the working of FDA in the United States only, European countries have also developed their own regulatory bodies like EMA and GDPR for the effective evaluation and organization of these AI-related technologies in the healthcare sectors. For successfully implementing the standards, the EU AI Act was passed in the European countries, rendering the industries to act according to the criteria set by the regulatory bodies for national and cross-border data management and transfer. The act was a risk-based and comprehensive part of jurisdictional landmarks that paved the way for governing the integration of AI in the healthcare sectors along with its development and deployment in terms of patient facilitation. This risk-based assessment approach was based on four different tiers of risk that were to be calculated while developing and deploying AI tools commercially. Minimal risk, limited risk, high risk, and unacceptable risk were the four types of risks that were instructed by the regulatory bodies in the usage of these AI. The minimal risk category encompasses such devices that pose little or no risk to patients and their related data, and a vast amount of AI software, such as spam filters, simple & complex types of video games, falls into this category. In comparison to this, software like chatbots (GPT, Gemini) that usually require the user to give their consent in using them, falls into limited risk category, as if the user is unwilling to give their credentials the system cannot do any harm. While the high-risk category is the most critical aspect of the EU AI Act as this encompasses such tools and software that are directly used in law enforcement agencies, infrastructure development, and the development of AI tools in the healthcare sectors (these machine and deep learning tools have a higher chance to interact with the life and death phases of the human beings). The last category is the unacceptable risk highlighting the clear threat of the AI tools in identification and screening of potential key metabolites, not only toward the patient's safety but also toward the industry itself. Techniques like government scoring and certain manipulative techniques fall into this category, and as per the act, those types of tools and software are strictly prohibited.

12.8.3 Nagoya Protocol on Access and Benefits Sharing (ABS)

Nagoya Protocol is a supplementary agreement (supplementary to the 1992 CBD) made in 2010 in Nagoya, Japan, to protect genetic biodiversity, and the main theme was access, distribution, equity, and utilization of genetic resources, ensuring sustainability of biodiversity across the globe. The protocol itself came into enforcement on 12 October 2014, when there was no compliance management conducted toward the users who were leaving or had left their country [63]. At present, a total of 139 countries have already been ratified by this protocol and is not just thought of as an international policy rather than an ethical contract, restricting the users from establishing fairness while harnessing natural products. The protocol came into existence because of several ethical concerns, and the main one being that many pharmaceutical sectors and big private healthcare firms were forcibly taking expensive medicinal plants without prior notice from the regulatory authorities. These firms were actively involved in screening, isolation, and development of medicinal drugs from the plant and were ripping off governments (keeping all the profits to themselves and were asking as much money as they could because of the products' scarcity), as the products made were novel and selective toward the treatment of a lot of internationally prevalent chronic illnesses. On the other hand, the countries or indigenous communities that were growing those novel plants were getting nothing in return. The ABS protocol was implemented to make sure anyone who is looking for certain indigenous plants for medicinal purposes must get the informed consent (PIC) from the host country and then should establish mutually agreed terms (MAT) by registering to the protocols of that specific country. To date, these restrictions have been effective in preventing the exploitation of biodiversity by giving no loose ends to biopiracy, which is a primary concern related to the unethical use of medicinal plants, as can be evident from the Amazonian tribes [64, 65].

As the use of AI models has been greatly influencing the trends in phytochemical industries by relying on pathway reconstruction, structured algorithms, predictive modeling, and virtual product screening and mining of metabolomes, the implementation of ABS is a dire need to ensure fairness and robustness alike. These AI models rely solely on big datasets of plant genomics, ethnopharmacological knowledge, and herb records, all of which fall under the categorization of the ABS protocol. Usage of such kind of mentioned data (from public databases) without prior notice may lead to the violation of protocols, even when the original source of that data is to be obtained from any sort of legal agreements. Countries are actively regulating the ABS rules in sorting and characterization of genomic datasets under the Nagoya Protocol, as this protocol provides the users with legal and ethical benefits to further enhance the phytochemical industries while protecting botanical biodiversity and indigenous communities that harbor novel plants.

12.8.4 United Kingdom (MHRA)

Apart from EMA, another regulatory body that works in the United Kingdom is the Medicines and Healthcare products Regulatory Agency (MHRA), whose purpose is similar to that of the FDA, GDPR, and EMA. The approach used by MHRA is based on AI assurance since the integration of any specific AI tool will be principle-driven, clear, concise, free of bias, and predictable in terms of its outcomes. The main theme is to create a check and balance on AI (SaMD devices) that must be used in high-risk areas in medical device design and manufacturing industry. A bundle of different domains was introduced by this regulatory authority to ensure the robustness of the protocols used in development and deployment of deep and machine learning tools used in phytochemical screening, and four of these technical domains are looking into model drift, data monitoring, algorithmic robustness, transparency, and accountability along with certain regulatory sandboxes to overcome the black-box nature of the AI tools [62]. Model drifting includes checking the degradation profile of the algorithm with time because of ever-changing real-world data, and to overcome the issue, the regulatory body emphasizes the use of continuous dataset monitoring to be effectively practiced by the respective sectors. Another challenge faced by the sectors was the lack of fairness in the ethical use of these tools, and MHRA put a strong check on mitigating the effects of bias. Prior to their commercialization, the developers must make sure that the presented model has been trained on a vast dataset (data from different demographics) to convey the fairness and generalization-like capabilities that will eventually help in diagnosis, treatment, or in the production or screening of novel drug compounds. Transparency and trust were the third domain on which the regulatory authority, MHRA, worked a lot. The sole purpose of this was to ensure the proper documentation, use of AI based on its Context of Use (COU), and the implementation of explainable AI (easy to understand Artificial Intelligence) that will ultimately help in maintaining trust and integrity of the patients toward the healthcare sectors, as they will be able to understand the working of these tools. Last but not the least, MHRA also introduced the use of regulatory sandboxes to closely monitor the developers whether they are ethically developing the software and tools or not and to halt the process of development at initial stages if something is against the protocols. The overall objective was to develop a framework that will assist the developers to check the product robustly at every stage of development and commercialization and so far, MHRA has proved beneficial [56, 66].

12.8.5 Asia-Pacific

While the regulatory framework was completely enforced in the United States and several countries of the United Kingdom, a lot of Asian countries also started to work on their regulatory programs to ensure the safety and integrity of AI-assisted tools in the medical industry. In China, National Medical Products Administration (NMPA) has begun to gain attention worldwide because of its strict protocols in implementing AI tools. The main objective of NMPA is to develop detailed, structured, and organized guidelines Performa for SaMD-like medical devices that are being used across the country to ensure the integrity of data quality and security, thus helping in the optimization of the AI products. Meanwhile, Japan has also established Ministry of Health, Labour and Welfare (MHLW) and Pharmaceuticals and Medical Devices Agency (PMDA) to regulate the working and implementation of AI-assisted medical devices (purposely used for screening and identification of plant-based metabolites) in pharmaceutical and healthcare sectors [58]. Continuous data improvement and validation of the model used are the two main domains on which the Japanese regulatory bodies put a strong cross-check on such AI-assisted devices that could adapt according to the given dataset on real-world evidence (RWE). While HSA (Health Sciences Authority) in Singapore has adopted a risk-proportionate approach for the acceleration of the already available AI-integrated devices in the country and the framework is error-free and mire agile than those of Japan and China, but still it is below the belt in comparison to the GDPR and

FDA. Iterative guidance is the key aspect of this regulatory body that helps maintain the fairness and safety of the machine learning model being used in the industry. In comparison to all those mentioned above, South Korea has also joined the race to maintain the safety and integrity of devices and machinery by forming the Ministry of Food and Drug Safety (MFDS) [67].

12.8.6 Other Global and Regional Policies

Apart from aforementioned universal drug regulatory policies, several other policies regarding the sole protection of plants and associated metabolites are also playing their role worldwide to ensure the protection of biodiversity. Convention on Biological Diversity (CBD) is the primary regulatory body, responsible for the protection and sustainability of biodiversity across all the international waters to ensure effective regulation of plant-based data. Another regulatory body, International Treaty on Plant Genetic Resources for Food and Agriculture (ITPGRFA), is also responsible for the fair trade of plant-based genetic materials across the world, ensuring transparency and fairness. WHO and WIPO traditional medical framework also protects the botanical knowledge under strict intellectual property rights and prevents misuse of biased and unethical commercialization of plant-based indigenous (novel) products. A lot of other regulatory bodies are also working; despite the US FDA in the United States, Dietary Supplement Health and Education Act (DSHEA, 1994) also regulates the commercialization of phytochemical-related products by integrating robust labeling and safety strategies. Furthermore, USDA Plant Protection Act also controls cross-border transfer of plant-based medicinal products to protect the prevention of pest and clonal diseases. Several other policies like UK Biodiversity Framework (United Kingdom), Biodiversity Conservation Strategy (China), China AI Regulations (2023-2024), Kampo Medicine Regulations (Japan), Herbal Medicine Safety laws (South Korea), and African Convention on the Conservation of Nature are working strictly in the protection of biodiversity, apart from the overall regulatory frameworks set by the FDA, MHRA and GDPR under the context of integration of AI tools in phytochemistry [68].

12.9 Exploring AI-Enabled Privacy and Security Through Real-World Cases

As the integration of AI tools in the pharmaceutical, phytochemistry (large-scale metabolic profiles, and bioactive prediction of active modalities) and healthcare sectors began to rise exponentially, the issue regarding their safety was also being critically evaluated as the cloud spaces (in which the data was to be stored) were in constant danger from the cyberattacks. Data tampering was one of the persistent threats that these models must face because the input was manipulated by the attackers which in turn will influence the expected results associated with the study being conducted. Surprisingly, it has been observed that the attacks against these trained models were taking different forms, but the main theme was the exact exploitation of the patient-related data. Another global security risk was the poisoning attack associated with the ethical use of tools, which makes learning difficult to interpret, and several defensive approaches are being utilized to cope with this issue as well, the one being the exclusion of the poisoned datasets from the study itself. One of the best approaches was the use of "data sanitization", ensuring the tainted samples that were to be removed from the dataset is to be trained on the deep or machine learning model. Detection and removal of unnecessary classifiers, along with dictionary attacks (by applying spam filters) from the error-prone databases, were the two main aspects of the data sanitization technique, and the sole purpose of these aspects was to protect the system from any sort of malicious attack by focusing on Reject on Negative Impact (RONI) [69]. After the successful implementation of data sanitization, new emerging sorts of poisoning attacks were also observed, and the attitude was shifted from the just identification of the tainted samples toward the development and deployment of such trained models that are resistant to any sort of poisoned cyberattacks. In case there is an attack done on the training model before its implementation and commercialization, there are several ways by which the data can be saved, and these include differential privacy, strengthening upgrades, and end-to-end encryption. To further strengthen the systems, the concepts of adversarial training were introduced by feeding the models with real-world tainted data to ensure their robustness so that adversarial scenarios could be created to understand the likelihood of the errors that may arise in the models. Although these adversarial training techniques are resistant to white-box attacks because of their ever-changing security, privacy, and robustness, still a lot more attention must be shifted toward the black-box attacks, as they are posing a significant threat making the algorithm vulnerable [70].

To completely understand how these poisoned attacks are constantly shaping the world's attitude toward the implementation of AI tools, real-world scenarios must be discussed to have a clearer picture. These scenarios will give us a clear and concise report on the success and failure of these trained machine learning models in the field of health. Meanwhile, the development and deployment of these tools are also studied to be affected by utilization of multinational approaches like organization, culture, regulatory standards, attention to detail, and continuous regulation. One best example of AI data breaches is the cyberattack on healthcare data that occurred in Japan, back in 2019 when a large US firm joined hands with one of the best industries to develop and commercialize an AI-assisted healthcare model. This act of friendship led to the unauthorized transfer of billions of patients' related sensitive information across the globe without the explicit user consent, leading to the rise of several questions on the robustness and ethical use of these devices [71]. Several researchers, scientists, and critics argue that although the sole purpose of this partnership was to develop something meaningful for the public, it neglected the key aspects of ethical and legal considerations like privacy, data protection, transparency, and informed consent. Meanwhile, another data breach occurred when National Health Service (NHS) joined hands with a private firm for the development of trained models. A total of 1.6 million patient-related information records were shared with the private industry in order to train a deep learning algorithmic model for the detection of kidney-related anomalies (main focus was on acute kidney injuries), but later the commissioner's office raised several questions on the data sharing and privacy, as patient consent was not justified before transferring the data to that specific industry [8, 53]. While the AI-based natural products are still posing less threat than the modern personalized medicines, the notable example is the misuse of these herbal products by Amazonian tribes without any proper consent from regulatory bodies. Another example was the case in which scientists risked the confidentiality of novel medicinal plants and utilized them for training deep and machine learning models. The issues related to the incident were the use of traditional knowledge without the benefit of sharing, in violation of Nagoya Protocols, and the leak of sensitive cultural information, which was openly accessed through AI. Another case was the unethical and biased mining of genomic data of Indian Ayurvedic plants that was stored within the online databases by various artificial intelligence agencies. Key concerns raised were the utilization of genomic data on an international scale without the consent of Biodiversity Authority of India. Another case of misuse of these phytochemicals was the case related with the information leaked from Southeast Asian Herbal Supplements as the researchers, unintentionally, classified a toxic compound as hepatotoxic alkaloids due to bias in training these models. Lack of transparency and compromised safety were the key concerns related to flawed decision-making by utilizing plant-based datasets [72].

In comparison to the abovementioned failures, several healthcare sectors have also implemented these tools and machines with a greater success rate while maintaining a solid cross-check on data security. One best example is the effectiveness of a telemedicine approach in Scandinavia, where the company utilized techniques like homomorphic encryption and data privacy by employing the approach known as Privacy-Preserving Machine Learning (PPML). Lately, the approach has also been strengthened by the integration of consent dashboards, third-party audits, and national transparency reports. Another notable example was the effective utilization of AI in a leading cancer research institute in Canada. The innovators designed an AI model for the prediction of administered dose-dependent treatment responses in patients by training the model on diverse genomic and hospital-related datasets from across Canada. The approach used in the development of this model was federated learning, as already mentioned, by excluding the sharing of raw data. The abovementioned examples show a clear picture of the advantages and potential ethical and legal risks associated with the use of these AI-assisted devices in the healthcare sector, highlighting major aspects such as the best infrastructure, privacy, transferability, and unbiased justifications can lead to better outcomes [34, 73].

12.10 Ethical and Societal Considerations

12.10.1 Ethical Use of Phytochemical Data

Apart from legal considerations in the effective use of these AI-driven models, ethical and societal considerations are also considered for understanding a range of concerns regarding their impact on public health and the values that they will bring to the healthcare sector. And for these reasons, a lot of internationally organized frameworks (already discussed in detail in the aforementioned context) have standardized the guidelines and protocols for the ethical use of these machines and deep learning algorithms. Autonomy, non-maleficence, beneficence, fairness, and justice are some of the key aspects regulated by the competent authorities in favor of the ethical use of AI machines. Respect for autonomy is by far the most foundational ethical concern, as patients must have an idea of how their personal and healthcare data is collected and processed in the medical sectors. Furthermore, non-maleficence (minimum or no threat to public health) is another concern that has aroused across national and international borders in response to the rising integration of corresponding algorithms in pharmaceutical industries. Beneficence, closely related to the respect for autonomy, is another ethical concern that emphasizes the use of these AI systems only to be used only in improving the patients' outcomes and safety. Last but not least, fairness and justice should also be kept in mind while working on these tools so that there is no chance of discrimination among the population and that the results are openly discussed with the underlying population or subjects. Regulatory authorities have issued a strong framework for the use of these tools in medical sciences because of the high-risk approach associated with these tools and their protocols must be constantly monitored and regulated to ensure the trained models are up to date with real-world data [8, 74].

12.10.2 Equity in AI-Enabled Phytochemistry

As these models have to be trained on large datasets (big data), problems can arise in data collection, data analysis, and data storage that will eventually disturb the expected outcomes and put a question mark on the predictability of the model, leading to unbiased data. Bias in training these AI-assisted datasets is another ethical concern, as most of the time, the researchers pick up data from a limited population (to minimize the cost and time of the experiment) that could lead to the breach of fairness, justice, and equity among the innovators and the subjects. A classic example is the use of these AI in translational oncology for the development and deployment of AI-assisted diagnostic and treatment plans in a transparent, fair, unbiased, and ubiquitous way that could be easily interpretable. One of the notable innovations, in terms of equity, is the use of federated learning (where models are trained on decentralized data in wearable IoT devices), enabling the model to be trained on data from various demographic, geographical, and institution-specific regions of the international community by negating the sensitive data transfer across these entities.

The working principle of the concept is solely based on preservation and security of patient-related data across healthcare industries while minimizing the integrity and robustness across heterogeneous societies (multiple groups of population). The technique is effective as it has been linked with the reduction of racial and ethnicity-based population in terms of translational oncology (diagnosis, screening, and treatment of cancer) by combining the data from multiple institutions all around the globe [74, 75].

12.10.3 Public Trust and Transparency

Transparency and public trust are the two most central ethical concerns in the context of Explainable AI (XAI), as the model, data, algorithms, and the results associated with these are to be openly discussed with the innovators, users, and the subjects alike. High-risk scenarios are to be openly discussed before training these models, and users must make sure the models trained are capable enough to calculate the risks with high chances of accuracy. Most of the time, these algorithms (trained model) give opaque results about a specific scenario, making it much more difficult for the user to understand the results (black-box). For that reason, ethical frameworks like human oversights were also introduced to make sure the proper functioning, such as support and professional judgment, and to focus on the ethical use rather than replacing the human force. Moreover, the issue of bias is still at large, as most of these models are trained on a limited dataset and may result in errors if some unseen data (related to different demographics) is incorporated by chance or intentionally. For example, if the dataset about the treatment efficiency (on a single population) of a certain registered medical product is to be fed to these models, they may give incomplete results that will not fit on the entire community and, in some rare cases, they may give negative results. To mitigate this issue, firms must ensure their professionals are trained on the technical aspects of dataset and model development. Furthermore, there is an urgent need to train these models on diverse datasets from different populations, following the specific guidelines set by the regulatory bodies [6–8].

12.11 Future Trends in Phytochemistry

The future of AI-assisted medical tools and devices in the healthcare sector is a global concern and is characterized by an international effort to put a strong emphasis and cross-check on the innovative approaches with ethical regulations and patients' compliance. Several key aspects like model governance, validation before deployment, and intellectual property rights are to be taken into account while discussing the AI integration. All the abovementioned aspects are still being discussed by and large under the four main contexts that in turn are interconnected as well and do include adaptive regulation, need to strengthen the IP law, standardization of protocols being used, and the implementation of regulatory sandboxes to overcome the problem of the black-box nature of the trained models.

12.11.1 Standardization of AI Validation

As discussed in the section on regulatory compliance, the current landscape is still under consistent threats, highlighting the fragmented nature of the regulations imposed by the governing bodies. The consistent differential strategy employed by US FDA and EU EMA is still under debate as the approach employed by the first one is flexible and dialogue-driven, while the latter one entirely focuses on the structured and risk-tiered nature of the AI algorithms used in the development and deployment of AI models. There is a dire need to develop a single regulatory framework, and bodies like WHO and ICH are continually working on developing a single, standardized model for the regulation of healthcare sectors to decrease the burden on the innovators and the users alike. The model will include several core concepts including the Good Machine Learning Practices (GMLP), the robustness of analytical approaches that will be used in data validation, clear and concise data representation, and other legal compliance to strengthen the overall model and ensure model consistency.

12.11.2 Continuous Monitoring and Adaptive Regulation

Another key aspect to look for is the continuous monitoring and adaptive regulation of the trained model to ensure the safety of the model itself and the safety of patient-related data. The main difference between the traditional regulation and AI-assisted continuous regulation and monitoring is the assumption of fixed products, but the latter systems greatly minimize the assumption of fixed products as this new model addresses post-marketing strategy in consideration with continuous monitoring. These mechanisms are also currently practiced by the US FDA (PCCP plan) and Japan (PACMP), and companies should pre-specify the modifications within algorithms along with adaptations of verification methods to ensure safety without compromising the regulatory review. The shift toward the abovementioned mechanisms highlighted the potential advantages of the Total Product Life Cycle (TPLC) over a one-time regulatory approach in making sure safety and efficacy are not compromised.

12.11.3 Regulatory Sandboxes and Target Pilot Approach

Sandboxes are the emerging tools that are effectively being utilized to overcome the vague, opaque nature of the model that is to be used in the diagnosis or treatment of typical diseases or in the screening and identification of novel drug moieties. The implementation of these models, as already done in the UK and EU pharmaceutical industries, will allow the innovators to examine the new AI-driven techs in an error-free and controlled environment encompassing real-world data under the consistent guidance of field experts. This controlled testing procedure will enable the users and industry workers to gain practical experience, understand the advantages and limitations of the trained model (to be prepared for any mishap), and redefine regulatory protocols set by the governing bodies to ensure the safety of the device prior to its commercialization. To further justify the legal practices, GDPR and HIPAA have also provided foundational rules for maintaining data integrity.

12.11.4 Strengthening IP Law for AI

Another challenge that was faced in implementing regulatory standards toward the ethical use of AI-assisted devices was the issue of intellectual property rights (who will be considered the owner), demanding a clear, robust, and more justified framework for IP protection. The strengthening of IP laws was further divided into two distinct categories: design patents and inventorship. AI inventorship emphasizes the justification of global policies toward the innovation of the trained model, highlighting the invention as a collaborative approach or a single entity to ensure the patentable nature of the novel moieties (drug targets). Molecular design of patents requires the regulatory bodies to set robust standards in screening, identification, and characterization of novel drug molecules. These approaches will further clarify the ways and means to protect the research and development (R&D) funding done by multinational biotech companies and to further minimize the risk factors associated with the FTO approach in the effective utilization of these AI models.

12.12 Conclusions

The main theme of this chapter is to highlight the key ethical and legal challenges that may arise in the development and screening of novel plant-based metabolites with the help of AI-assisted models (deep and machine learning models). As AI is progressively enhancing the pharmaceutical sector by enabling the identification, screening, characterization, development, and commercialization of potential phytochemicals that will assist in targeted therapy by enhancing pharmacovigilance and pharmacoepidemiologic studies. The integration of real-world data (RWE) into the utilization of AI-assisted models is helping human beings in several attempts, but the rising concerns regarding the legal and ethical challenges associated with these still demand a protected, robust, and collaborative framework across all the international platforms to minimize the risks associated with their use. To date, the foundational aspects of the AI industry are potentially under consistent threat due to several unresolved legal considerations, discussed in detail in the aforementioned context, like lack of proper licensing and data ownership. Data protection and privacy are non-negotiable as per the protocols set by the GDPR for fair, unbiased, lawful data processing and by HIPAA regulations set to protect patient-related data from potential cybersecurity breaches. Moreover, the lack of understanding of the re-identification risks in the localization of data (as seen in China) and a lack of clear policy for FTO analysis in correspondence with AI inventorship still remain at large. To overcome all the issues associated with ethical and legal compliance, regulatory bodies like FDA, EME, MHRA, GDPR, HIPAA, and the EU AI Act are advising the healthcare industries to practice GMLPs in order to enforce risk-based classificational approaches, predetermined control plans, transparency, accountability, accessibility, cross-border transport, safety, reduction of bias, AI model validation, use of XAI, utilization of regulatory sandboxes, data equity, data security, data confidentiality, and fairness to tackle the black-box nature of AI algorithms.

References

1. DiMasi, J. A., Grabowski, H. G., & Hansen, R. W. (2016). Innovation in the pharmaceutical industry: New estimates of R&D costs. *Journal of Health Economics, 47*, 20–33.
2. ElSohly, M. A., et al. (2017). Phytochemistry of *Cannabis sativa* L. *Progress in the Chemistry of Organic Natural Products, 103*, 1–36.
3. Banu, K. S., & Cathrine, L. (2015). General techniques involved in phytochemical analysis. *International Journal of Advanced Research in Chemical Science (IJARCS), 2*(4), 25–32.
4. Opderbeck, D. W. (2019). Artificial intelligence in pharmaceuticals, biologics, and medical devices: Present and future regulatory models. *Fordham Law Review, 88*, 553.
5. WHO Guidance. (2021). *Ethics and governance of artificial intelligence for health*. WHO.
6. Gerke, S., Minssen, T., & Cohen, G. (2020). Ethical and legal challenges of artificial intelligence-driven healthcare. In *Artificial intelligence in healthcare* (pp. 295–336). Elsevier.
7. Ravali, R. S., et al. (2023). Pioneering ethical boundaries: Empowering AI governance for the future of pharmaceutical and nursing sectors. *Latin American Journal of Pharmacy, 42*(3), 695–701.
8. Shaki, F., et al. (2024). Artificial intelligence in pharmaceuticals: Exploring applications and legal challenges. *Pharmaceutical and Biomedical Research, 10*(1), 1–10.
9. Muehlematter, U. J., Daniore, P., & Vokinger, K. N. (2021). Approval of artificial intelligence and machine learning-based medical devices in the USA and Europe (2015–20): A comparative analysis. *The Lancet Digital Health, 3*(3), e195–e203.
10. Warraich, H. J., Tazbaz, T., & Califf, R. M. (2025). FDA perspective on the regulation of artificial intelligence in health care and biomedicine. *JAMA, 333*(3), 241–247.
11. Wu, C., et al. (2020). Scientific and regulatory approach to botanical drug development: A US FDA perspective. *Journal of Natural Products, 83*(2), 552–562.
12. Lenarczyk, G., et al. (2025). The future of AI regulation in drug development: A comparative analysis. *Journal of Law and the Biosciences, 12*(2), lsaf028.
13. Vamathevan, J., et al. (2019). Applications of machine learning in drug discovery and development. *Nature Reviews. Drug Discovery, 18*(6), 463–477.
14. Wang, L., et al. (2019). Artificial intelligence facilitates drug design in the big data era. *Chemometrics and Intelligent Laboratory Systems, 194*, 103850.
15. Blanco-Gonzalez, A., et al. (2023). The role of AI in drug discovery: Challenges, opportunities, and strategies. *Pharmaceuticals, 16*(6), 891.
16. Scheeder, C., Heigwer, F., & Boutros, M. (2018). Machine learning and image-based profiling in drug discovery. *Current Opinion in Systems Biology, 10*, 43–52.
17. Ary, T., & S.N.T. (2019). Celebrating 25 years of DSHEA. *WholeFoods Magazine, 42*, 20–45.
18. Gilbert, N. (2011). Regulations: Herbal medicine rule book. *Nature, 480*(7378), S98–S99.

19. Chagas-Paula, D. A., et al. (2015). Prediction of anti-inflammatory plants and discovery of their biomarkers by machine learning algorithms and metabolomic studies. *Planta Medica, 81*(06), 450–458.
20. Wang, L., et al. (2021). Establishment of a lung cancer discriminative model based on an optimized support vector machine algorithm and study of key targets of Wogonin in lung cancer. *Frontiers in Pharmacology, 12*, 728937.
21. Varghese, R., et al. (2025). Artificial intelligence driven approaches in phytochemical research: Trends and prospects. *Phytochemistry Reviews, 24*, 1 16.
22. Lee, K. B., et al. (2018). Phytochemical profiles of Brassicaceae vegetables and their multivariate characterization using chemometrics. *Applied Biological Chemistry, 61*(2), 131–144.
23. Cunha-Oliveira, T., Ioannidis, J. P., & Oliveira, P. J. (2024). Best practices for data management and sharing in experimental biomedical research. *Physiological Reviews, 104*(3), 1387–1408.
24. Echendu, U., & Udeokechukwu, C. (2025). *Artificial intelligence and machine learning in healthcare: Developing privacy-preserving frameworks*. Available at SSRN: https://doi.org/10.2139/ssrn.5125847
25. Schaduangrat, N., et al. (2020). Towards reproducible computational drug discovery. *Journal of Cheminformatics, 12*(1), 9.
26. Nature, E. J. (2021). Replicating scientific results is tough—But crucial. *Nature, 600*(7888), 359–360.
27. Council, N.R, et al. (2002). *Integrity in scientific research: Creating an environment that promotes responsible conduct*. National Academies Press.
28. Carbon, S., et al. (2019). An analysis and metric of reusable data licensing practices for biomedical resources. *PLoS ONE, 14*(3), e0213090.
29. Grabus, S., & Greenberg, J. (2019). The landscape of rights and licensing initiatives for data sharing. *Data Science Journal, 18*, 29–29.
30. Benjamin, W. (2020). Data localization and ASEAN economic community. *Asian Journal of International Law, 10*(1), 158–180.
31. Al-Worafi, Y. M. (2020). Herbal medicines safety issues. In *Drug safety in developing countries* (pp. 163–178). Elsevier.
32. Taneja, H., & Singh, A. K. (2015). Preserving privacy of patients based on re-identification risk. *Procedia Computer Science, 70*, 448–454.
33. Tsuji, S., et al. (2021). Artificial intelligence-based computational framework for drug-target prioritization and inference of novel repositionable drugs for Alzheimer's disease. *Alzheimer's Research & Therapy, 13*(1), 92.
34. Gawankar, S., et al. (2024). Patient privacy and data security in the era of AI-driven healthcare. In *2024 8th international conference on computing, communication, control and automation (ICCUBEA)*. IEEE.
35. Li, H., Yu, L., & He, W. (2019). *The impact of GDPR on global technology development* (pp. 1–6). Taylor & Francis.
36. Dave, A., & Agrawal, A. K. (2025). A comparative study with GDPR, HIPAA, CCPA, PIPEDA and DPDPA. In *2025 IEEE International Conference on Computer, Electronics, Electrical Engineering & Their Applications (IC2E3)*. IEEE.
37. Mazzi, F. (2022). *Patentability of AI-generated inventions: A case study on pharma*. Maastricht University/Queen Mary University of London.
38. Comer, A. C. (2020). AI: Artificial inventor or the real deal? *NCJL & Technology, 22*, 447.
39. Poddar, A., & Rao, S. R. (2024). Evolving intellectual property landscape for AI-driven innovations in the biomedical sector: Opportunities in stable IP regime for shared success. *Frontiers in Artificial Intelligence, 7*, 1372161.
40. Otto, S. (2021). An approach to the de novo synthesis of life. *Accounts of Chemical Research, 55*(2), 145–155.
41. Visan, A. I., & Negut, I. J. L. (2024). Integrating artificial intelligence for drug discovery in the context of revolutionizing drug delivery. *Life (Basel), 14*(2), 233.

42. Mohd Shith Putera, N. S. F., et al. (2024). Bridging legal education and practice: An empirical insights into artificial intelligence accountability in healthcare. *Asian Journal of University Education, 20*(3), 790–806.
43. Arrieta, A. B., et al. (2020). Explainable Artificial Intelligence (XAI): Concepts, taxonomies, opportunities and challenges toward responsible AI. *Information Fusion, 58*, 82–115.
44. Kırboğa, K. K., et al. (2023). Explainability and white box in drug discovery. *Chemical Biology & Drug Design, 102*(1), 217–233.
45. Gilbert, S., et al. (2021). Algorithm change protocols in the regulation of adaptive machine learning–based medical devices. *Journal of Medical Internet Research, 23*(10), e30545.
46. Rezaei, S., & Liu, X. (2019). Deep learning for encrypted traffic classification: An overview. *IEEE Communications Magazine, 5*(5), 76–81.
47. Mintz, Y., & Brodie, R. (2019). Introduction to artificial intelligence in medicine. *Minimally Invasive Therapy & Allied Technologies, 28*(2), 73–81.
48. Massella, M., et al. (2022). Regulatory considerations on the use of machine learning based tools in clinical trials. *Health Technology (Berlin), 12*(6), 1085–1096.
49. Solomon, D. H., & Rudin, R. S. (2020). Digital health technologies: Opportunities and challenges in rheumatology. *Nature Reviews Rheumatology, 16*(9), 525–535.
50. Bresso, E., et al. (2021). Investigating ADR mechanisms with knowledge graph mining and explainable AI: A feasibility study with knowledge graph mining. *BMC Medical Informatics and Decision Making, 21*(1), 171.
51. Chen, B. K., & Yang, Y. T. (2013). Post-marketing surveillance of prescription drug safety: Past, present, and future. *The Journal of Legal Medicine, 34*(2), 193–213.
52. Payrovnaziri, S. N., et al. (2020). Explainable artificial intelligence models using real-world electronic health record data: A systematic scoping review. *Journal of the American Medical Informatics Association, 27*(7), 1173–1185.
53. Martini, N., et al. (2020). Expert opinion on Real-World Evidence (RWE) in drug development and usage. *Pharmadvances, 2*(2), 41–50.
54. Van Batavia, J., & Goldenberg, S. (2019). *FDA Device Regulation: 510 (k), PMA*. ITMAT Education.
55. Act, E. A. I. (2024). *The EU Artificial Intelligence Act*. European Union.
56. Papapetropoulos, A., et al. (2024). Novel drugs approved by the EMA, the FDA, and the MHRA in 2023: A year in review. *British Journal of Pharmacology, 181*(11), 1553–1575.
57. Zheng, L., et al. (2024). Targeted drug approvals in 2023: Breakthroughs by the FDA and NMPA. *Signal Transduction and Targeted Therapy, 9*(1), 46.
58. Pollard, S. (2022). Considerations in Heading East: A beginner's guide to Japan and PMDA. *Orthopedics and Design Technology, 18*(6), 20–22.
59. Fasshauer, S. (2023). Regulatory landscape in Singapore. In *Handbook of cell and gene therapy* (pp. 243–254). CRC Press.
60. Azer, Ö. A., & Oran, İ. B. (2023). AI applications in health sector: Use of artificial intelligence in Covid-19 crisis and impacts of medical robots on global economy. In *Corporate sustainability in times of virus crises* (pp. 203–224). Springer.
61. Rome, B. N., Kramer, D. B., & Kesselheim, A. S. (2014). FDA approval of cardiac implantable electronic devices via original and supplement premarket approval pathways, 1979-2012. *JAMA, 311*(4), 385–391.
62. Ryu, G., & Lee, J. (2021). Comparison of Classification Rules Regarding SaMD Between the Regulation EU 2017/745 and the Directive 93/42/EEC. *Journal of Biomedical Engineering Research, 42*(6), 277–286.
63. Morinaka, Y., & Takamura, Y. (2017). Recent developments in Japanese implementation of the Nagoya Protocol. *Journal of International Relations and Area Studies, 46*, 13–23.
64. Nijar, G. S., et al. (2017). The implementation of the Nagoya ABS Protocol for the research sector: Experience and challenges. *International Environmental Agreements: Politics, Law and Economics, 17*(5), 607–621.
65. Martins, J., et al. (2020). The Nagoya Protocol and its implications on the EU Atlantic area countries. *Journal of Marine Science and Engineering, 8*(2), 92.

66. Smith, D., et al. (2017). Explanation of the Nagoya Protocol on access and benefit sharing and its implication for microbiology. *Microbiology, 163*(3), 289–296.
67. Lim, K., Heo, T.-Y., & Yun, J. (2022). Trends in the approval and quality management of artificial intelligence medical devices in the Republic of Korea. *Diagnostics, 12*(2), 355.
68. Chihomvu, P., et al. (2024). Phytochemicals in drug discovery—A confluence of tradition and innovation. *International Journal of Molecular Sciences, 25*(16), 8792.
69. Carmody, J., et al. (2021). AI and privacy concerns: A smart meter case study. *Journal of Information, Communication and Ethics in Society, 19*(4), 492–505.
70. Nguyen, T. M., et al. (2022). Explaining black box drug target prediction through model agnostic counterfactual samples. *IEEE/ACM Transactions on Computational Biology and Bioinformatics, 20*(2), 1020–1029.
71. Kikuchi, H., et al. (2021). Best security measures to reduce cyber-incident and data breach risks. In *International Workshop on Data Privacy Management*. Springer.
72. Rastogi, S., et al. (2007). Adverse effects of Ayurvedic drugs: An overview of causes and possibilities in reference to a case of Vatsanabha (Aconite) overdosing. *The International Journal of Risk & Safety in Medicine, 19*(3), 117–125.
73. Gadade, D. D., et al. (2024). Pushing boundaries: The landscape of AI-Driven drug discovery and development with insights into regulatory aspects. In *Artificial intelligence and machine learning in drug design and development* (pp. 533–561). Wiley-Scrivener.
74. Deo, R. C. (2024). Artificial intelligence and health equity. *JACC Advances, 3*(9). https://doi.org/10.1016/j.jacadv.2024.101045
75. Srivastav, A. K., et al. (2025). Revolutionizing oncology through AI: Addressing cancer disparities by improving screening, treatment, and survival outcomes via integration of social determinants of health. *Cancers, 17*(17), 2866.

Managing Security Risks in AI Deployment

13

Nimra Mudassir, Asim Waris, and Adeeb Shehzad

Abstract

Artificial intelligence (AI) systems introduce a rapidly evolving security landscape in which conventional cybersecurity approaches are no longer sufficient. Because AI applications are data-dependent, statistically driven, and continuously updated, they expose novel attack surfaces including data poisoning, backdoor insertion, model inversion and extraction, adversarial perturbations, and privacy-compromising inference attacks. This chapter proposes a structured taxonomy of these risks across data, model, infrastructure, human governance, and societal levels, demonstrating how adversaries exploit vulnerabilities throughout the AI lifecycle from compromised data pipelines and unverified training sources to misconfigured cloud deployment and insecure model supply chains. Existing governance frameworks, including the European Union AI Act, the National Institute of Standards and Technology (NIST) AI Risk Management Framework, and the OECD AI Principles, reflect increasing regulatory convergence yet lack harmonized operational security requirements and standardized evaluation metrics. To address these gaps, this chapter presents a Secure AI Development Lifecycle (AI-SDLC) that incorporates risk-aware design, trustworthy data governance, adversarial robust model training, secure MLOps deployment, and post-deployment monitoring. Empirical case studies across healthcare, autonomous

N. Mudassir · A. Waris
Department of Biomedical Sciences and Engineering, School of Mechanical and Manufacturing Engineering, National University of Science and Technology, Islamabad, Pakistan

A. Shehzad (✉)
Biodiversity Research Center, Dhofar University, Salalah, Oman

Y. Bibi et al. (eds.), *Artificial Intelligence in Phytochemistry*,
https://doi.org/10.1007/978-981-92-0875-3_13

systems, financial platforms, and pharmaceutical R&D illustrate the real-world consequences of insecure AI deployment. This chapter concludes by emphasizing the imperative for defense-in-depth security, sustained human oversight in high-risk environments, and international cooperation to ensure that AI technologies remain trustworthy, safe, and aligned with public welfare.

Keywords

Artificial intelligence · Cybersecurity · AI security landscape · Data poisoning · Backdoor insertion

13.1 Introduction

Artificial intelligence (AI) refers to computational systems that can perceive their environment, learn from data, and perform tasks that normally require human intelligence, such as reasoning, decision-making, and language understanding. In its broadest sense, AI encompasses rule-based expert systems, machine learning (ML) algorithms and more recent deep neural networks [1]. The deployment of AI systems is the process of integrating trained models into operational environments so that they interact with users or other systems to produce predictions, recommendations, or automated actions. Deployment is not merely a technical step; it also involves integration with organizational workflows, regulatory compliance, and continuous monitoring.

Security risks in AI encompass threats to the confidentiality, integrity, and availability of both training and operational data, the models themselves and the infrastructure that surrounds them. Data can be manipulated to poison the learning process [2], models can be extracted by adversaries [3], and inputs can be subtly altered to produce erroneous outputs [3, 4]. These risks differ from classic information security because they exploit statistical learning mechanisms rather than exploiting conventional software vulnerabilities. For example, the lack of human oversight and automation of data collection has exposed organizations to attacks in which malicious samples are injected into training data to control downstream behavior [5]. Adversaries can also perform membership or model inversion attacks to recover sensitive training examples [6].

Traditional IT security measures are insufficient for AI because the attack surfaces extend into the data, the learning algorithms, and the statistical properties of the models. The integrity of a model can be compromised without ever accessing the underlying code; adversaries can query a model-as-a-service and approximate its decision boundaries to steal its functionality [3]. High-confidence misclassifications are possible due to small perturbations to inputs, and this can compromise the safety of medical diagnostic models [4]. Additionally, AI systems can be vulnerable to both unintended and intentional harm because they usually use large and heterogeneous datasets that can be biased or harbor sensitive information. Such problems require an integrated multi-layered security strategy that cuts across data management, model resilience, and infrastructural security.

The increasing recognition of AI societal, ethical, and security impacts has accelerated regulatory action across jurisdictions, with governments adopting proactive frameworks to mitigate systemic and high-risk AI harms. European Union AI Act follows a risk-driven approach, which is also extraterritorial; any provider or deployer, whose AI output is utilized within the EU shall bear the responsibility to meet tough conditions [7]. It categorizes AI systems as prohibited, high-risk, and limited-risk categories and places the requirements of risk management systems, documentation, and post-market surveillance. In the United States, the National Institute of Standards and Technology (NIST) issued the AI Risk Management Framework (AI RMF) that outlines governance, mapping, measurement, and management of functions to deal with AI risks [8]. The other global values like Organization for Economic Co-operation and Development (OECD) AI Principles emphasize fairness, transparency, and accountability. Collectively, these frameworks are an indication of the fact that voluntary guidelines will be replaced with binding standards, and strong security and governance of AI applications are necessary.

This chapter presents a theoretical and practical situation of the security threat in AI implementation. It integrates the available literature about data, model, and infrastructure threats as well as details technical and organizational controls required to address the threats. Through comparative analysis of regulatory frameworks and relying on case studies, this chapter will help practitioners, researchers, and policymakers to create secure and trustful systems based on AI. The sections look at a particular part of the AI security environment in terms of taxonomy of threats, and the weakness of the traditional cybersecurity controls up to new risk management frameworks and considerations.

13.2 Conceptual Foundations

13.2.1 Evolution of AI Security

The field of artificial intelligence security has developed into a niche academic topic but has since become a major concern of the field of computer science and public policy. Prior studies of adversarial machine learning have shown that neural networks whose behavior has been trained on apparently innocent input can be ruined by well-constructed perturbations or poisonous input, which results in misclassifications and service failure [9, 10]. Researchers soon realized that these attacks are not the same as typical software exploits: they begin exploiting the statistical characteristics of learning algorithms as opposed to the weaknesses present in code. The prevalence of data poisoning, backdoors, and model extraction, where adversaries can inject their illicit samples during training or affect third party datasets, was later recorded [11]. The domain broadened with the systematic surveys mapping the proliferation of adversarial methods and defenses to image, text, and sensor-based systems [9]. Simultaneous efforts on the software supply chains demonstrated how open-source packages and pre-trained models can introduce malicious code, resulting in the downstream compromise [12]. All these studies together established the conceptual framework of AI security and provided an opportunity to emphasize the necessity of specific threat models that help to represent the lifecycle of data and models instead of considering only the running code.

13.2.2 Differences Between AI Security and Traditional Cybersecurity

The inherent peculiarities of AI systems, including non-determinism, dynamic adaption, and data dependency, generate attack surfaces which classical information systems do not have. In contrast to traditional software, where vulnerabilities are a result of logic and/or configuration errors, AI models are trained to learn; thus, they have whatever biases, errors, or malicious code contained in their data sets. The opponents can contaminate a dataset to drive an autonomous vehicle or a medical model into making perilous choices [11]. Sensitive information can also be recovered by attackers through trained models through membership inference or model inversion [6, 13]. This reliance on third-party data and models also further expands the attack surface; vulnerable sensors can feed vulnerabilities through an organization pipeline, or corrupted open-source libraries can spread vulnerabilities down the pipeline [11]. Such facts imply that traditional security measures in the form of antivirus programs or firewalls are no longer adequate [9]. Moreover, AI risk assessment needs to deal with dynamic and uncertain probabilities and ethical aspects; risk management scholars observe that there are no recommended ways of estimating probability or impacts of cyber-physical attacks on complex IoT systems [14]. The field hence demands new measurements, interdisciplinary experience, and constant examination throughout the training, deployment, and operational stages.

13.2.3 AI System Lifecycle and Threat Modeling

The AI system consists of several processes problem formulation, data collection, preprocessing, model training, validation, deployment, and post-deployment monitoring. The stages bring in distinct assets and possible ways of failure. As a case in point, the data can be obtained through clinical records, online repositories, or sensors; a contamination at this point will spread across the entire pipeline [11]. During training, model parameters can be manipulated via backdoors or poisoned labels, and even after deployment, adversaries can craft inputs to trigger misclassifications [10]. To systematically analyze these risks, researchers have adapted classical threat-modeling methods. The STRIDE-AI methodology maps failure modes to specific threats such as spoofing, tampering, and information disclosure across AI assets [11]. It combines Failure Mode and Effects Analysis (FMEA) with an asset-centered view, helping practitioners prioritize controls for data, model, and infrastructure artifacts. Other work stresses the lack of agreed-upon standards for AI threat modeling and calls for lifecycle-wide approaches that include data quality

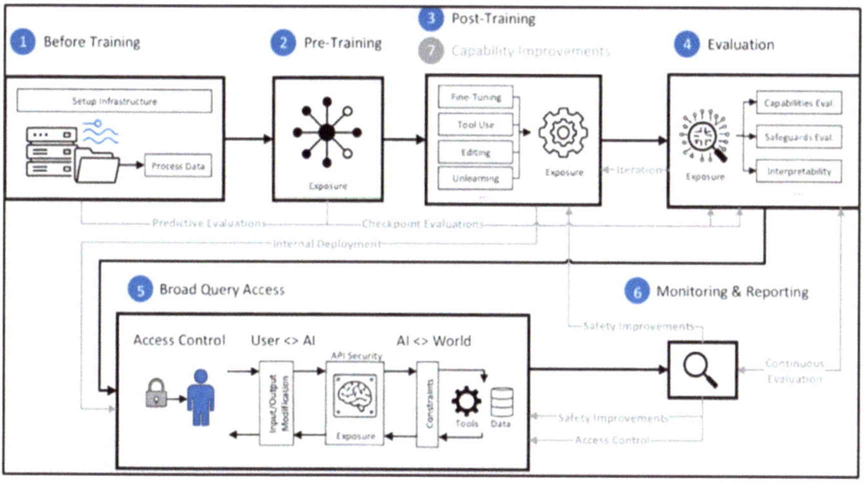

Fig. 13.1 AI Security Lifecycle (AI-SDLC). (Adapted from Vassilev et al. [15])

audits, adversarial robustness testing, and supply-chain security [14]. Understanding these conceptual underpinnings is essential before examining the taxonomy of AI security risks in detail (Fig. 13.1).

13.3 Taxonomy of Security Risks in AI Deployment

13.3.1 Data- and Model-Related Risks

The most direct threats to AI deployment concern the data that feed models and the models themselves. Poisoning attacks occur when adversaries deliberately insert malicious samples into training data to influence the learned decision boundary. For instance, a recent study showed that online misinformation could corrupt the large language models used in medical diagnosis, causing them to produce harmful recommendations [16]. Because many AI systems are trained on web-scraped corpora such as Common Crawl and Open Web Text, they are exposed to incidental contamination as well; medical concepts extracted from these datasets are particularly vulnerable [16]. Beyond poisoning, attackers may exploit insecure data pipelines to tamper with data in transit or at rest, circumventing integrity checks.

In laboratory-to-model workflows, the risk extends to the provenance of chemical and biological data. Biospecimen metadata, chromatographic spectral characteristics, and assay conditions can be altered to sabotage compound-screening outcomes [17]. When such manipulated records pass into machine-learning pipelines undetected, downstream target-prediction models inherit these erroneous assumptions, resulting in costly false positives or masking of toxicity signals. Ensuring secure data lineage through cryptographic signing and automated

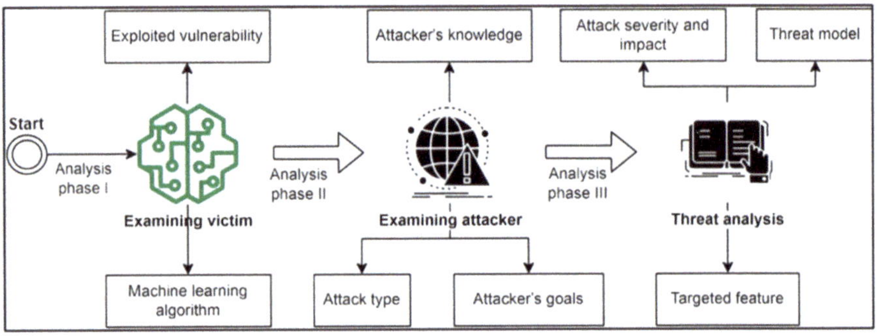

Fig. 13.2 Threat-analysis process for machine-learning systems. (Adapted from Paracha et al. [45])

quality-control checks is therefore a core requirement in pharmaceutical AI deployments [18, 19].

Confidentiality is also at stake: membership inference and model inversion attacks allow adversaries to determine whether a particular individual's data was included in training or to reconstruct sensitive features from the model's outputs. Differential privacy mechanisms can mitigate these risks, but even privacy-preserving training is not immune to adaptive attacks [13] (Fig. 13.2).

Model-level threats are equally diverse. Model stealing and extraction techniques enable adversaries to approximate or replicate proprietary models by querying them repeatedly [3]. Adversarial examples, imperceptible perturbations crafted to mislead classifiers, can cause safety-critical systems, such as medical imaging algorithms, to make high-confidence errors [4]. Backdoor and Trojan attacks embed triggers within a model during training; when the trigger appears at inference time, the model outputs a malicious label [20]. Such backdoors may persist across model updates and can be difficult to detect without specialized defenses. Finally, models evolve after deployment: distributional shift and data drift can gradually erode performance and create unforeseen safety hazards [21]. Persistent monitoring and regular retraining are therefore essential components of a security strategy.

AI-driven drug discovery pipelines are particularly susceptible to data poisoning because biological assay datasets are commonly aggregated from heterogeneous laboratory processes and crowdsourced compound-screening repositories. Malicious manipulation of molecular descriptors, experimental labels, or docking scores can cause models to prioritize inactive or toxic candidates while suppressing genuinely active molecules. Attackers can exploit this vulnerability to derail discovery timelines or intentionally introduce harmful compounds into downstream development [22]. In addition, the confidentiality of training data in pharmaceutical environments intensifies the risk of membership inference and property-inference attacks, where adversaries attempt to uncover proprietary formulations, structural fingerprints, or pre-clinical data embedded in learned representations [23].

Recent evaluations demonstrate that adversarial molecular perturbations subtle changes to atoms, charges, or stereochemistry can cause activity-prediction networks to misclassify highly toxic compounds as safe drug leads [24]. These attacks pose dual-layer harm: they compromise scientific integrity and increase the likelihood of real-world patient exposure to dangerous substances. Industry studies argue that pharmaceutical AI is becoming a target of industrial espionage, where attackers leverage property-inference attacks to identify commercial drug scaffolds earlier than competitors, undermining innovation and regulatory exclusivity (Table 13.1).

To mitigate dual-use outcomes in molecular design, security-informed defensive strategies are increasingly being embedded into generative pipelines. Constraint-based generation enforces medicinal-chemistry safety rules such as Lipinski's filters, PAINS (Pan-Assay Interference Compounds) exclusion lists, and toxicity prediction gates, preventing harmful scaffolds from being proposed. Moreover, formal verification techniques and kill-switch mechanisms are being explored to ensure that generated structures remain within approved therapeutic classes and cannot be steered toward weaponizable compounds. These defensive controls must be continuously updated as adversaries devise new perturbation strategies to bypass biochemical safety rules.

Table 13.1 Common AI tools used in phytochemistry and drug discovery and associated security risks [24–27]

No.	AI tool type	Example application in phytochemistry/drug discovery	Primary security risk
1	Machine Learning Classifiers	Bioactivity prediction, QSAR modeling	Data poisoning attacks exploiting label manipulation
2	Generative AI Models (e.g., VAE, GAN, Diffusion Models)	De novo molecular generation	Dual-use misuse for designing harmful compounds
3	Graph Neural Networks	Structure–activity and binding affinity prediction	Model theft and IP leakage via query extraction
4	Deep Learning Spectral Analysis	NMR, MS, FTIR structural interpretation	Adversarial perturbation affecting peak recognition
5	Reinforcement Learning Agents	Molecular optimization and synthesis planning	Reward-hacking leading to unstable or toxic molecules
6	Large Language Models (LLMs)	Plant-compound–target knowledge extraction	Hallucination risks and sensitive data leakage

13.3.2 Infrastructure, Operational, and Human Governance Risks

AI systems do not exist in isolation; they rely on complex infrastructures and human operators. Supply-chain vulnerabilities in software and model dependencies can introduce malicious code or vulnerabilities downstream. Studies of software supply chains reveal that selective attacks spread through multi-layer networks more rapidly than random attacks and that improving software quality and security investment is more effective at mitigating them [28]. Open-source ecosystems are particularly vulnerable because malicious packages can be distributed through trusted repositories such as npm or PyPI; high-profile incidents like Not Petya and the compromised CCleaner update demonstrate the reach of such attacks. When the AI services are deployed, they reveal API endpoints that can be listed and abused; the lack of proper integration of the AI services with legacy systems or improperly configured cloud environments further increases the attack surface. Pipelines on the basis of continuous-integration/continuous-delivery can choose to include contaminated components unwittingly, and a weak identity and access control system can permit unauthorized users to modify training jobs or model artifacts [12].

There are specific difficulties with insider threats (either malicious or negligent) when the operators can access sensitive training data or model weights. The untrained personnel can change systems or turn off security measures in order to attain better performance. Campaigns based on social engineering against data-annotation personnel or machine-learning engineers can possibly result in credential compromise, allowing adversaries to inject poisoned information or steal models. Furthermore, the organizations do not always have a well-defined system of governance over AI: they might not have a human-in-the-loop to make high-impact decisions, may not separate responsibilities between the development and security teams, and may not adequately oversee third-party vendors [29]. These areas of weaknesses demonstrate the significance of adopting security into the operations of MLOps and create responsibility in AI implementation.

13.3.3 Systemic, Ethical, and Societal Risks

Outside technical and operational risks, there are system-wide problems, which involve ethics, fairness, and social welfare. Other risks in AI systems include bias in training data when historical injustices or sampling bias are used; in the medical sector, biased models may provide different predictions to various groups of patients, which increases inequality. The slogan bias in, bias out underlines that biased data lead to biased results, and bias may be injected into the model lifecycle at any point. Explainability, the lack of which is also a feature of many AI systems, makes it difficult to interpret or challenge any decisions made by the model, negatively affecting trust and responsibility. Too much automation may destroy human experience and enable subtle mistakes to spread out unnoticed. Cascading failures are another issue: since AI systems are usually operationalized in sensitive systems, vulnerabilities in one area of AI may lead to failures in other interdependent systems. These risks have national security implications as well; adversarial attacks on autonomous vehicles or PowerGrid optimization algorithms could disrupt transport or energy supply. Addressing systemic risks therefore requires integrating ethical review boards, fairness audits, and societal impact assessments into the development process [30].

13.4 Why Are Traditional Security Measures Not Enough

13.4.1 Limitations of Conventional Cybersecurity Controls

AI systems expose weaknesses in conventional security controls. Antivirus software, firewalls, and intrusion detection systems are designed to block known malware signatures or abnormal network traffic. They are ill suited to detect adversarial examples or data-poisoning attacks that exploit the statistical nature of learning models. Quantitative risk assessments used in information security often compute expected loss as likelihood times consequence; however, for emerging cyber-physical systems such as the Internet of Things, there is scant data on attack probabilities or impacts, making such assessments unreliable. Researchers argue that there are no standards or policies governing IoT cyber-risk measurement and that the real impact of attacks is unknown [14]. Traditional defenses also assume that software vulnerabilities can be patched once discovered, but in AI systems vulnerabilities may stem from data biases or model design choices that persist even after code updates [10]. These shortcomings illustrate why AI security cannot be approached as a variant of classical cybersecurity.

13.4.2 Unique Vulnerabilities in AI Pipelines and Attack Surfaces

AI pipelines introduce novel attack surfaces that conventional security tools do not monitor. Attackers can poison training data or plant a backdoors prior to a model even being deployed [11]. Since organizations receive datasets or pre-trained models through third parties, infected components can bypass supply chains. There are ways of manipulating sensors and physical devices that submit data to AI systems to provide adversarial inputs that elude digital controls. Unlike normal software vulnerabilities, these attacks take advantage of dynamic nature of learning not of the fixed code defects. Existing threat-modeling tools are not applicable to AI assets; one study observes that there is no well-developed solution to consider threats at the full machine-learning lifecycle [11]. Consequently, organizations may fail to identify tampering or information disclosure risks embedded in data pipelines or model artifacts until after deployment.

13.4.3 Black-Box Models and Lack of Mature AI Security Standards

Many AI models operate as black boxes: their internal representations are opaque even to their creators. This opacity hampers security audits and makes it difficult to detect whether a model has been backdoored or stolen. Scholars have expressed concern that current AI security research focuses on ingenious attack and defense mechanisms without developing clear guidance on secure development practices. Cyber-risk in emerging technologies is further complicated by dynamic system interdependencies and cascading risks in IoT networks [14]. Bias can be introduced at any stage of the AI lifecycle, yet existing security standards rarely address how to audit for bias or ensure fairness [30]. Moreover, there are not widely adopted, sector-agnostic standards for AI security comparable to ISO 27001 in information security. Policymakers have begun drafting regulations such as the EU AI Act and the NIST AI Risk Management Framework, but these remain high-level and do not substitute for technical controls. As a result, organizations must develop their own risk management strategies and adapt existing frameworks to account for AI's unique characteristics.

13.5 AI Risk Management Frameworks

13.5.1 Global and Industry Frameworks

The growing recognition of AI's societal impact has spurred the development of risk management frameworks by governments, standards bodies, and industry consortia. In the United States, the National Institute of Standards and Technology (NIST) released the AI Risk Management Framework 1.0 in 2023. The framework outlines four core functions Govern, Map, Measure, and Manage to help organizations incorporate trustworthiness considerations into the design, development, and deployment of AI [8]. The European Union's AI Act classifies systems into prohibited, high-risk, and limited-risk categories and imposes obligations such as risk management systems, data quality requirements, documentation, and post-market surveillance [7]. International guidelines such as the Organization for Economic Co-operation and Development (OECD AI Principles focus on human rights, transparency, robustness, and accountability [8]. Complementing these public initiatives are emerging industry frameworks: Google's Secure AI Framework (SAIF) and Microsoft's Secure AI Framework provide checklists for secure development and deployment, while OpenAI publishes risk controls for large language models. The standard ISO/IEC 42001:2023 presents an Artificial Intelligence Management System standard, which mandates organizations to implement policies, procedures, and controls throughout the AI lifecycle, despite limited formal literature on ISO 42001, the uptake of it by larger cloud providers is an indicator of a direction toward certifiable AI governance. All these structures are indicative of a change in voluntary principles to enforceable guidelines, which will generate incentives on safe AI activities within industries.

13.5.2 Comparative Analysis and Gaps

The frameworks differ in scope, prescriptiveness, and maturity, although they have common goals. The AI RMF offered by NIST is not compulsory and focuses on risk profiling and stakeholder involvement, which offers flexibility to industries of various kinds [8]. The EU AI Act is legally enforceable in the EU and has extraterritorial effects requiring providers and users of high-risk AI systems to adopt quality management systems and conformity assessments [7]. OECD principles, on the other hand, are soft law, which provides broad principles that have to be converted into sector-based policies [8]. Industry frameworks, such as Google's Secure AI Framework (SAIF), guidelines published by Microsoft and risk controls offered by OpenAI, provide practical checklists, but fail to tackle the systemic problems of bias, human control, or supply-chain integrity. Furthermore, current models have been criticized on their failure to consider human factors and failure to have some measures used to gauge trustworthiness [1]. The level of integration between cybersecurity requirements (e.g., ISO 27001) and AI-specific requirements is also limited. The lack of standardised frameworks makes it difficult to evaluate compliance and benchmarking performance by organizations, prompting calls for harmonization across jurisdictions and sectors [11]. With the emergence of new regulations, such as the proposed AI Bill of Rights in the United States, and industry-specific guidance in healthcare services and finance practitioners must navigate an increasingly complex landscape of regulatory requirements. Such comparative analysis, therefore, shows that step forward to formalize AI governance, as well as gaps that persist and require further studies and coordination.

13.6 Secure AI Development Lifecycle (AI-SDLC)

13.6.1 Secure Design and Data Governance

Secure design and good data governance are the initial steps to building secure AI systems. Requirements engineering should incorporate threat modeling in order to recognize when the data, models, and artifacts may fail and rank controls. Strategies such as STRIDE-AI and FMEA, which map failure modes to threats throughout the lifecycle, with the focus on data poisoning, backdoors, and supply-chain tampering, should be discussed during the design phase. Secure data pipelines require provenance tracking, validation, and sanitization to prevent contaminated or malicious samples from entering training sets [11]. Federated learning and differing privacy are privacy-preserving techniques that help avoid sensitive data violating privacy and provide an option to use collaborative learning; recent surveys reveal that federated learning, differential privacy, and their combinations are considered the most common approach to medical data protection [10]. Clear governance policies for data collection, labeling, and storage mitigate the risk of bias and ensure compliance with regulations.

13.6.2 Secure Model Training and Evaluation

Data collection, labeling, and storage have clear policies of governance that alleviate the chances of bias and adherence to regulations. Adversarial training is one of the most effective defense mechanisms, whereby models are trained on adversarial examples; this method is highly recommended in IoT and health applications [9]. Perturbation resistance can be further improved by using robust optimization and feature-squeezing methods. Membership inference is a privacy attack that needs to be tuned and monitored by using differential privacy budgets [13]. Identifiers can be added to models using watermarking and fingerprinting in order to detect illegal duplication or manipulation [20]. Secure validation is done by testing out-of-distribution and adversarial input models to reveal any hidden vulnerabilities; red-team exercises need to model realistic attacker capabilities both in white-box and black-box environments [10]. These activities help to ensure a model satisfies safety and performance.

13.6.3 Secure Deployment, MLOps Controls, and Post-Deployment Monitoring

There are additional risks that are brought about by the deployment stage whereby models are incorporated into working systems. MLOps services are to apply robust identity and access control, role-based access control, and secure API gateways to avert unauthorized accessibility [29]. Pipelines based on continuous-integration/continuous-delivery have to scan dependencies and prevent malicious code [28]. Network segmentation, zero-trust, and cloud security controls are examples of infrastructure hardening measures that minimize the blast radius of intrusions and lateral movement. Once deployed, data drift and degradation of performance in models should be monitored; such minor changes in inputs distributions can lead to various clinical or financial models to generate an incorrect diagnosis or classification [21]. Retraining or rollback procedures should be triggered by alerts, and model-specific forensics should be covered by incident response plans. Considering the dynamicity of AI, post-deployment monitoring is a process rather than a single activity and a foundation of safe and trustworthy operation [14].

13.7 Technical and Operational Controls

13.7.1 Data Security Controls

Trustworthy artificial intelligence is based on data security. Differential privacy adds controlled noise to training or inference to ensure that the chance of any specific sample being differentiated is limited and hence membership inference and model inversion attacks are alleviated [13]. Federated learning also enables two or more organizations to jointly train models without exchanging raw information which lessens exposure to centralized breaches [10]. Those techniques may be used together: federated learning updates can be subjected to the differential privacy in order to reduce leakage even further [13]. Two additional methods are secure multi-party computation, in which cryptographic protocols are used to compute the model gradients without revealing the input data, and synthetic data generation, in which artificial datasets are generated with the same statistical properties while preserving the privacy of sensitive records. Although synthetic data has the potential to minimize privacy risks, it still can reveal rare or sensitive patterns that are not audited carefully [10]. Strong data governance also involves strong ingestion pipelines by encrypting them and performing integrity and access control to prevent tampering [11]. Lastly, companies should put certain data-retention and data-deletion policies in place to reduce the span of vulnerability and adhere to the laws like the General Data Protection Regulation (GDPR) and the Health Insurance Portability and Accountability Act (HIPAA) [10].

13.7.2 Model Hardening and Infrastructure Protection

In addition to data, models and the infrastructure that surrounds them have to be hardened. The adversarial training method is one of the major defense mechanisms where the model is exposed to adversarial examples in the course of training, and thus it learns stronger boundaries of decisions [9]. Certified defense and distributionally robust training are powerful optimization methods that can offer provable protection against some types of attacks. Watermarking and fingerprinting implant marks on model weights or outputs to assert ownership and unauthorized copying [20]. Before they are passed to the model, input validation and sanitization, e.g., rejecting out-of-range or suspicious queries, can prevent injection and evasion attacks. From an infrastructure perspective, zero-trust systems and network segmentation restrict the sideways movement and isolate sensitive elements. The secure API gateways can be used to support authentication, rate limiting, and request validation to prevent the abuse of model endpoints. The identity and access management should uphold the least-privilege principles such that training data or model artifacts can only be accessed by the authorized personnel and services [29]. Environments in clouds are supposed to be developed with encryption, logging, and runtime monitoring that can identify anomalies and aid in incident response [14]. Lastly, the supply-chain security mechanisms, which include scanning dependencies of malicious code and signatures on pre-trained models, are needed to avoid upstream compromises [28].

13.7.3 Threat Detection and Incident Response in AI Systems

Threat detection and incident response are essential for securing AI systems, as attacks like data poisoning, adversarial inputs, and model extraction often appear as subtle anomalies rather than explicit failure. Continuous monitoring of data pipelines and model behaviour is therefore required to identify deviations in input distributions, prediction patterns, and system performance. Techniques such as anomaly detection and data drift analysis can help detect potential threats at an early stage. These detection mechanisms provide the foundation for structured monitoring systems and enable timely response strategies to mitigate risks in AI depolyment.

13.7.4 Monitoring and Detection

Active surveillance is required to prevent the threats before they can cause unimaginable damage. The existence of data poisoning can be manifested by the continuous monitoring of data pipelines and model behavior which might show some unexpected changes in feature distributions or unusual correlations [16]. Due to the inclusion of heterogeneous web-scraped data in many AI systems, monitoring should also go further than internal datasets and encompass the provenance and integrity of the third-party data. Inference and model inversion attacks present statistical artifacts like abnormal confidence scores or query patterns which can be detected using anomaly detection algorithms [13] Data drift detection algorithms, like the observation of input statistics and output calibration, can identify when models are being used with shifted distributions and can no longer be trusted. Model performance and fairness metrics surveillance can reveal the emerging bias or performance degradation [30]. All these monitoring methods must be combined together in automated dashboards which should alert operators of suspicious trends and lead to additional research.

13.7.5 Incident Response and Recovery

Organizations should respond as soon as an attack is realized to limit damage and ensure that trust is regained. Playbooks that are specific to AI incident response must describe roles and duties, malware isolation model, and data breach mitigation. In fact, patient harm occurred due to a ransomware attack on a hospital in Germany since the IT system was not redundant and had no contingency plans [29]. It needs to be responded to using the cooperation of data scientists, security, and domain specialists who will conduct root-cause analysis, such as forensic inspection of training data, model weights, and logs. Supply-chain attacks can require the rejection of trust in third-party models and dependency list updates [28]. Models that were found to have backdoors are to be retrained to clean data or fixed by methods like subnetwork replacement [20]. Lesson learned must be captured in incident reports and input into risk registers and threat models [11]. The analysis of the incident after it has occurred can expose the systemic vulnerabilities like poor training of staff or improperly configured access controls [29]. Recovery plans should involve some communication with impacted stakeholders, reporting and constant monitoring of the regulations to eradicate possible re-exploitation [14].

13.8 Governance, Policies, and Organizational Controls

13.8.1 Human Oversight, Ethics, and Documentation

The problem of security in AI is not only a technical challenge, but, in the first place, it is a question of government and organizational culture. High-impact decisions cannot be made without human attention, so that there is the ability to challenge an automated recommendation and correct it. Research on cyber-incidents in healthcare reveals that insufficient employee training, poor password, and maintenance are some of the factors contributing to breach and loss of lives of patients. Proposed AI deployments need to be examined by ethical review boards and interdisciplinary committees on whether they involve any bias, discriminatory effects, and ethical issues [30]. This governance involves documentation. Data provenance refers to standardized documentation describing data sources, model architecture, training objectives, limitations, and intended use, often presented through artefacts as datasheets for datasets, model cards, and system cards, which allow auditors and regulators to assess risk. Bias detection frameworks emphasize that developers of AI, domain experts, and security professionals can all be responsible in introducing bias at any point in the AI lifecycle and that multi-faceted and holistic solutions are necessary to reduce it [29, 30]. In addition to that, human-in-the-loop mechanism can help to avoid automation bias and make sure that decisions are made in accordance with the organizational values and regulations.

13.8.2 Compliance Requirements and Vendor Risk Management

The management of risks associated with vendors and third-party elements and attempting to cope with a complex environment of regulations is also a part of AI governance. The EU AI Act puts risk-based duties on the providers and users and demands the high-risk systems to adopt quality management, risk management, and post-market surveillance [7]. Healthcare organizations will also have to adhere to sectoral laws including HIPAA in the United States and the General Data Protection Regulation in the EU which introduce stringent requirements on the collection, storage, and processing of personal information [10]. The NIST AI Risk Management Framework provides voluntary guidance on governance, yet it underlines the importance of mapping risks, measuring risks, and controlling risks [8]. The third-party and vendor risk management is especially difficult, as the development of AI is usually based on open-source libraries, pre-trained models, and cloud services. The studies of the software supply chains have revealed that deliberate attacks on multi-layer networks are fast spreading and that investing in better software quality and security is more effective than patching the vulnerabilities only [28]. Evil people have taken advantage of open-source ecosystems and smuggle Trojan code into popular packages [12]. Vendors should be vetted harshly by organizations, transparency on training data and model provenance should be required, and dependencies ought to be monitored in terms of vulnerability. Contracts should include clauses on security obligations, incident reporting, and access to audit logs. Finally, internal policies must align with evolving regulations and industry standards to ensure continuous compliance and to avoid legal and reputational repercussions [29].

13.9 Case Studies of AI Security Failures

13.9.1 Real-World AI Security Failures in Critical Domains

Empirical examples illustrate how AI security lapses can translate into tangible harm. In one early poisoning attack on image classification models used for biomedical data, researchers injected mislabeled images into the training set, causing systematic misdiagnosis of diabetic retinopathy [2]. More recently, a study of medical large language models demonstrated that introducing misinformation during training led the models to recommend harmful treatments or ignore contraindications [16]. Adversarial attacks on autonomous vehicles show that minor alterations to street signs or lane markings can cause vehicles to swerve or brake suddenly [4]. Model extraction attacks against public APIs including conversational agents have reproduced proprietary models with only query access [3], undermining intellectual property and enabling further attacks. Data drift has also led to real-world harm: a healthcare model trained on pre-pandemic data failed to recognize atypical presentations of disease during the COVID-19 outbreak, leading to misdiagnoses and delayed care [21]. Supply-chain attacks, such as the insertion of malware into widely used software libraries (e.g., NotPetya and the compromised CCleaner update), have indirectly affected AI systems that rely on these libraries [12].

13.9.2 Lessons Learned and Cross-Domain Patterns

These case studies share common themes. First, inadequate data governance exposes AI systems to poisoning and drift; collecting data from untrusted or dynamic sources without rigorous validation allows malicious or outdated information to propagate into models [16]. Second, lack of robust training and validation enables adversarial and backdoor attacks; defensive techniques such as adversarial training, input sanitization, and model watermarking were absent in many failures [20]. Third, organizations also underestimated how easily attackers may extract or duplicate models using public interfaces, and this may require hardening of APIs and monitoring their usage [3]. Fourth, real-life events simultaneously depict that AI failures are rarely single events; they have to interact with system weaknesses, such as supply-chain attacks and human factors. The misdiagnosis instances highlight the necessity of constant observation and drift identification in the high risk areas [21]. Lastly, inter-domain patterns indicate that multiple-layered controls are needed such as secure data pipelines, strong model training, hardened infrastructure, and organizational governance. The lessons of these failures guide the best practices and teach that security in AI is a process and not a certification.

13.10 Sector-Specific Security Considerations

13.10.1 Healthcare

The medical industry is possibly the most sensitive field to apply AI, since any malfunction may directly harm the patients. The AI systems in this case have to defend very sensitive personal information and adhere to the regulations like the HIPAA and GDPR [10]. There are nursing records of the fact that hospital networks are appealing to hackers who may obtain unauthorized access to electronic health records and manipulate the clinical process or sell the data on the black market. Medical devices that are run using AI (insulin pumps, ventilators, etc.) can be remotely updated or disrupted by a ransomware attack, and in one case, a German hospital resulted in the death of patients after devices shut down due to the attack. These risks are compounded by system-level vulnerabilities that are outdated software, poor interoperability, and poor staff training [29]. Bias in healthcare data can also be propagated by AI models, which can result in different treatment recommendations [30]. In this case, data drift can lead to the misclassification of new disease patterns by diagnostic models, and this is what happened during the COVID-19 pandemic. Good security would thus involve good encryption, rigorous access, adversarial training, fairness auditing, perpetual drift, and bias monitoring.

13.10.2 Finance and Banking

In finance, AI systems underpin credit scoring, fraud detection, and algorithmic trading. Attackers may extract proprietary models from public APIs to build competing services or to evade fraud detection [3]. Membership inference and model inversion attacks can reveal whether an individual's financial data were used for training and potentially reconstruct sensitive attributes [13]. Biased training data can lead to discriminatory lending practices, raising regulatory and reputational concerns. Moreover, adversarial examples designed to mimic legitimate transactions can bypass anomaly detection systems. Financial institutions must therefore implement strong API authentication, rate limiting, model watermarking, and privacy-preserving training. Regulatory frameworks such as the EU AI Act and sector-specific guidelines from banking authorities are beginning to address these issues but remain in early stages of adoption.

13.10.3 Public Health, Defense, and Critical Infrastructure

AI is increasingly embedded in public health surveillance, national defense, and critical infrastructure management. In smart cities, IoT devices and ML models coordinate traffic, energy, and emergency services. Adversarial attacks on these systems could misdirect autonomous vehicles or disrupt power grids [9]. Cyber-risk researchers highlight that new technologies introduce risks that traditional assessment methods have not anticipated and that there is a lack of standardized metrics for evaluating IoT cyber-risk. Supply-chain vulnerabilities in software dependencies present a serious threat: malicious code inserted into open-source packages can propagate across infrastructure [28]. In national security contexts, AI systems may be targeted by state-sponsored actors seeking to extract models or insert backdoors. Given the interdependencies among sectors, attacks on one critical system can cascade into others. Mitigation strategies include network segmentation, zero-trust architectures, continuous threat modeling, and collaboration between public agencies and private vendors. Standards for critical infrastructure (e.g., NIST's Cybersecurity Framework) should be extended to explicitly cover AI components and supply-chain risks [14].

13.10.4 Pharmaceutical R&D and Drug Discovery Security

AI systems in pharmaceutical and natural-product drug discovery face high-impact security risks because adversarial manipulation can misclassify hazardous compounds as promising candidates, compromise scientific integrity, and create serious biosafety hazards. Subtle perturbations to molecular descriptors or chemical fingerprints can bypass toxicity-filtering constraints, allowing compounds structurally similar to organophosphate nerve-agent analogues to be ranked as top drug leads and advance into costly wet-lab validation stages without detection [31]. Attackers who infiltrate automated pipelines such as active-learning screening systems, cloud-integrated docking workflows, or robotic synthesis platforms can alter experimental outputs, disrupt LIMS-based traceability, or leak proprietary chemical structures for commercial sabotage or malicious synthesis. Ensuring compliance with Good Laboratory Practice (GLP) and Good Manufacturing Practice (GMP) therefore requires secure data lineage, toxicity-aware adversarial testing, and human-in-the-loop oversight to prevent dangerous, unintended, or illegal outcomes in AI-enabled drug discovery environments [32].

13.11 Future Directions in AI Security

As AI systems increasingly automate laboratory and pharmaceutical workflows, new biosecurity challenges are emerging. Autonomous AI agents capable of controlling robotic experiments or molecular-generation pipelines may be manipulated to design or synthesize harmful compounds. Continuous red-teaming, safe-reinforcement learning, and kill-switch mechanisms are therefore essential to ensure fail-safe behavior in safety-critical environments.

Global harmonization of standards is also needed, because vulnerabilities in one jurisdiction can rapidly propagate across interconnected pharmaceutical supply chains. Shared threat-intelligence networks and coordinated regulatory oversight will play a central role in preventing cascading failures that could endanger public health and national security.

13.11.1 Evolving Red-Teaming and Responsible Scaling

Defenders must keep pace with adversaries. Sophisticated attackers no longer simply craft imperceptible perturbations; they can steal a model's functionality by repeatedly querying an API and using the responses to infer its architecture and decision boundaries, ultimately constructing a substitute model [33]. These model-extraction attacks threaten intellectual-property rights and enable fraud across domains such as language models and autonomous vehicles [33]. Attackers can also embed backdoors during training so that a specific trigger causes the model to misbehave; NIST notes that such backdoor attacks are unique because they require the attacker to control both training and testing data, unlike other attacks that only target one stage of the pipeline [15]. To remain effective, red-teams must therefore emulate a broad range of attacker capabilities white-box, gray-box, and black-box and exercise control across the entire AI pipeline. Modern research emphasizes automated adversarial generation; for instance, universal adversarial perturbations are input-agnostic noise vectors that can cause misclassification across many inputs, and generator-based methods use generative adversarial networks to craft such perturbations [34]. Responsible scaling policies should accompany these testing efforts; unrestrained deployment of powerful models without rigorous evaluation can magnify harms, particularly when open-access models can be extracted and repurposed by adversaries [33]. The Block Center's 2024 workshop on red-teaming underscores that the US executive order on AI safety calls for NIST to develop rigorous red-team testing guidelines before public release [35], highlighting the need for structured simulations that go beyond prompting for harmful outputs and incorporate model extraction, backdoor insertion, and data-poisoning scenarios. Future work should integrate red-teaming into continuous integration pipelines, develop red-team maturity models, and encourage sharing of best practices across sectors.

13.11.2 Autonomous AI Agents and Alignment with Safety Engineering

The rise of autonomous AI agents' systems that make decisions and act in the world with little or no human oversight creates new security challenges. In industrial and off-highway vehicles, functional safety standards such as IEC 61508 and ISO 13849 ensure machines fail safely when components malfunction; however, Safety of the

Intended Functionality (SOTIF) emphasizes that systems must also behave safely when components work correctly but encounter performance limitations (e.g., sensors misinterpreting shadows or dust) [36]. As autonomy increases, both functional safety (FuSa) and SOTIF are essential [36]. Key standards, including IEC 61508, ISO 13849, IEC 62061, and ISO 19014, require manufacturers to demonstrate that safety circuits can reliably cut power or control movement even if an electronic component fails. Future regulations, including the EU Machinery Regulation (2023/1230) which are instigated in 2027, will require the specific coverage of AI and changing logic; a machine utilizing AI or machine learning, in spite of change in behavior following aftermarket release, must stay safe. They also state that system-level verification of safety components and demonstration of the fact that systems do not cause hazards during normal operation is required, not only when they fail [36].

Safety engineering principles should therefore be embedded in AI agents. The AI Safety, Ethics and Society textbook lists key safe-design features including redundancy, separation of duties, least privilege, fail-safes, antifragility, negative-feedback mechanisms, transparency, and defense in depth to ensure systems behave predictably and limit harm when things go wrong [37]. Human oversight remains indispensable. In healthcare, guidelines ask whether staff are trained to recognize when manual intervention is needed and whether clear protocols exist for overriding AI recommendations [38]. Alignment research must also address ethical issues: bias can be introduced at every stage of the AI lifecycle from conceptualization and data collection through algorithm development, implementation, and surveillance so robust bias detection frameworks are needed [30]. Autonomous agents will need to reason about long-term consequences, adapt to dynamic environments, and withstand adversarial inputs; recent work proposes formalized safe reinforcement-learning methods that integrate model checking and spatiotemporal rule verification to guide reward design and monitor decisions, producing safer state action strategies [39]. Such advances in safe reinforcement learning and formal verification help ensure agents behave within specified safety constraints.

13.11.3 Global Harmonization and Integration with Safety Standards

The Organizations operating across jurisdictions face divergent AI regulations and guidelines. The EU's Artificial Intelligence Act classifies AI systems into risk categories: unacceptable-risk systems (e.g., manipulative social-scoring or subliminal AI) are prohibited, high-risk systems are regulated, limited-risk systems face transparency obligations, and minimal-risk systems are unregulated [40]. Providers of high-risk AI must establish risk-management systems, ensure training and testing data are representative and free from errors, create technical documentation, design systems for event logging, and allow human oversight [40]. On the contrary, the US NIST AI Risk Management Framework is voluntary and is meant to offer a flexible and rights saving guideline in various sectors [41]. The OECD AI Principles promote transparency, accountability, and resilient and secure AI all through the lifecycle [42] and are non-binding [43]. The proliferation of the models may cause existing confusion and non-uniform security practices. Researchers point out that the lack of standardized measures to assess AI systems particularly those which incorporate logic and visibility of agent interactions discourages system-wide observability and risk management [44].

The harmonization initiatives must, therefore, incorporate AI-specific controls into a set of standards of safety and cybersecurity. Functional safety standards (IEC 61508, ISO 13849, IEC 62061, ISO 19014) are applied to safety critical systems in domains such as off highway vehicles so the system is capable of detecting faults and transitioning to a safe state, have redundancy in critical systems and predictable behavior under any fault condition [36]. These standards are complemented by best practice frameworks like ISO 21448 (SOTIF), ISO/PAS 8800 (safety of AI/ML in road vehicles), UL 4600 (autonomous product safety), ISO/IEC TR 5469:2022 (applying functional safety principles to AI systems), and ISO/IEC 42001:2023 (AI management systems) [36]. The threat intelligence and incident reports also require international collaboration; model extraction surveys indicate that vulnerabilities are found not in individual language models, autonomous vehicles, edge computing, and federated learning [33], but in the system as a whole. With AI systems becoming part of the critical infrastructure, metrics, certification schemes, and interoperability should be coordinated globally.

13.12 Conclusion

Securing AI requires a *multi-layered strategy* addressing technical, organizational, and societal risks. Traditional software protections fail against AI-specific threats such as data poisoning, model inversion, extraction, and backdoors. Effective defense demands secure data pipelines, adversarially hardened training, privacy-preserving methods, robust deployment, and continuous monitoring. Regulatory frameworks like the *EU AI Act* and *NIST AI RMF* emphasize risk-based governance, while industry promotes secure design and observation. Human oversight, transparency, and harmonized standards are vital to mitigate bias, drift, and systemic failures. Case studies in healthcare, finance, and supply chains show the consequences of weak security and the benefits of layered controls. Ultimately, embedding *security-by-design*, ensuring accountability, and fostering international collaboration will safeguard AI systems, protect public trust, and enable safe innovation.

References

1. Polemi, N., et al. (2024). Challenges and efforts in managing AI trustworthiness risks: A state of knowledge. *Frontiers in Big Data, 7*, 1381163.
2. Mozaffari-Kermani, M., et al. (2014). Systematic poisoning attacks on and defenses for machine learning in healthcare. *IEEE Journal of Biomedical and Health Informatics, 19*(6), 1893–1905.

3. Liu, Y., et al. (2023). Shrewdattack: Low cost high accuracy model extraction. *Entropy, 25*(2), 282.
4. Finlayson, S. G., et al. (2019). Adversarial attacks on medical machine learning. *Science, 363*(6433), 1287–1289.
5. Goldblum, M., et al. (2022). Dataset security for machine learning: Data poisoning, backdoor attacks, and defenses. *IEEE Transactions on Pattern Analysis and Machine Intelligence, 45*(2), 1563–1580.
6. Veale, M., Binns, R., & Edwards, L. (2018). Algorithms that remember: Model inversion attacks and data protection law. *Philosophical Transactions of the Royal Society A: Mathematical, Physical and Engineering Sciences, 376*(2133), 20180083.
7. Busch, F., et al. (2024). Navigating the European Union artificial intelligence act for healthcare. *npj Digital Medicine, 7*(1), 210.
8. Makridis, C. A., et al. (2024). From theory to practice: Harmonizing taxonomies of trustworthy AI. *Health Policy Open, 7*, 100128.
9. Harbi, Y., et al. (2024). Roadmap of adversarial machine learning in internet of things-enabled security systems. *Sensors (Basel), 24*(16), 5150.
10. de Aguiar, E. J., Traina, C., Jr., & Traina, A. J. (2023). Security and privacy in machine learning for health systems: Strategies and challenges. *Yearbook of Medical Informatics, 32*(01), 269–281.
11. Mauri, L., & Damiani, E. (2022). Modeling threats to AI-ML systems using STRIDE. *Sensors, 22*(17), 6662.
12. Ohm, M., et al. (2020). Backstabber's knife collection: A review of open source software supply chain attacks. In *International conference on detection of intrusions and malware, and vulnerability assessment*. Springer.
13. Riaz, S., et al. (2023). Membership inference attack on differentially private block coordinate descent. *PeerJ Computer Science, 9*, e1616.
14. Radanliev, P., et al. (2024). AI security and cyber risk in IoT systems. *Frontiers in Big Data, 7*, 1402745.
15. Vassilev, A., et al. (2024). *Adversarial machine learning*. NIST.
16. Alber, D. A., et al. (2025). Medical large language models are vulnerable to data-poisoning attacks. *Nature Medicine, 31*(2), 618–626.
17. Berens, P., et al. (2023). AI for science: An emerging agenda. *arXiv preprint*, arXiv:2303.04217. https://doi.org/10.48550/arXiv.2303.04217
18. Ballester, P. J. (2019). Machine learning for molecular modelling in drug design. *Biomolecules, 9*, 216.
19. James, P. (2025). *AI-driven risk assessment models in concurrent biopharmaceutical development projects*.
20. Wang, H., et al. (2022). Trap and replace: Defending backdoor attacks by trapping them into an easy-to-replace subnetwork. *Advances in Neural Information Processing Systems, 35*, 36026–36039.
21. Sahiner, B., et al. (2023). Data drift in medical machine learning: Implications and potential remedies. *The British Journal of Radiology, 96*(1150), 20220878.
22. Chen, W., et al. (2023). Artificial intelligence for drug discovery: Resources, methods, and applications. *Molecular Therapy Nucleic Acids, 31*, 691–702.
23. Gheorghita, F.-I., Bocanet, V.-I., & Iantovics, L. B. (2025). Machine learning-based drug-drug interaction prediction: A critical review of models, limitations, and data challenges. *Frontiers in Pharmacology, 16*, 1632775.
24. Zhan, H., et al. (2024). Accelerated pure shift NMR spectroscopy with deep learning. *Analytical Chemistry, 96*(4), 1515–1521.
25. Ghosh, S., et al. (2025). Computational advancements to facilitate therapeutic application of phytochemicals: Where do we stand? *Discover Applied Sciences, 7*(5), 491.
26. Walters, W. P., & Barzilay, R. (2021). Critical assessment of AI in drug discovery. *Expert Opinion on Drug Discovery, 16*(9), 937–947.
27. Liu, J., et al. (2024). Vibrational spectroscopy can be vulnerable to adversarial attacks. *Analytical Chemistry, 96*(42), 16570–16580.

28. Cai, H., Xiong, Q., & Lian, S. (2025). Research on network security vulnerability risk contagion in software supply chain based on system dynamics. *PLoS One, 20*(11), e0335128.
29. Di Palma, G., et al. (2025). AI-induced cybersecurity risks in healthcare: A narrative review of blockchain-based solutions within a clinical risk management framework. *Risk Management and Healthcare Policy, 18*, 3479–3497.
30. Hasanzadeh, F., et al. (2025). Bias recognition and mitigation strategies in artificial intelligence healthcare applications. *npj Digital Medicine, 8*(1), 154.
31. Ambery, P., Kilcoyne, A., & O'Connor, D. (2025). *Pharmaceutical medicine*. Oxford University Press.
32. Bahari, N. G. M., et al. (2018). The role of artificial intelligence in drug discovery: Accelerating the development of new medicines. *International Journal of Health Sciences, 2*(1), 64–78.
33. Zhao, K., et al. (2025). A systematic survey of model extraction attacks and defenses: State-of-the-art and perspectives. *arXiv preprint*, arXiv:2508.15031. https://doi.org/10.48550/arXiv.2508.15031
34. Sun, B., Sun, J., & Zhao, W. (2025). Democratic training against universal adversarial perturbations. *arXiv preprint*, arXiv:2502.05542. https://doi.org/10.48550/arXiv.2502.05542
35. Sinha, A., et al. (2025). *What can generative AI red-teaming learn from cyber red-teaming?* (Technical Report CMU/SEI-2025-TR-006). Software Engineering Institute.
36. Krishnamurthy, P. (2024). *Digital twin system for automotive safety with integration of a safety controller* (Master thesis).
37. Hendrycks, D. (2025). *Introduction to AI safety, ethics, and society*. Taylor & Francis.
38. Bignami, E., et al. (2025). AI policy in healthcare: A checklist-based methodology for structured implementation. *Journal of Anesthesia, Analgesia and Critical Care, 5*(1), 56.
39. Yin, C., et al. (2023). Spatio-temporal rule constraint guided safe reinforcement learning for CPS. In *2023 10th International conference on dependable systems and their applications (DSA)* (pp. 455–465). IEEE.
40. Ferri, F. (2025). The artificial intelligence act: An overview. In *An introduction to digital data law in the EU* (p. 96). Giappichelli.
41. NIST. (2023). *Artificial intelligence risk management framework (AI RMF 1.0)*, p. 100–1. https://nvlpubs.nist.gov/nistpubs/ai/nist.ai
42. OECD. (2023). *OECD digital economy papers*.
43. Bashir, H., & Ravi, M. (2025). A systematic review of strategic artificial intelligence governance frameworks and mechanisms in studies from 2017-2025. *Virtual space and social media studies, 2*(3), 9–54.
44. Coimbatore Ramalingam, B. (2025). Efficient implementation of AI agents in Enterprise application integration (EAI) and electronic data interchange (EDI). *International Journal of Scientific Research in Computer Science, Engineering and Information Technology, 11*(2), 10.32628.
45. Paracha, A., et al. (2024). Machine learning security and privacy: A review of threats and countermeasures. *EURASIP Journal on Information Security, 2024*(1), 10.

Strategic Solutions for Overcoming Challenges in AI Adoption

14

Tehreem Baig, Muhammad Jawad Khan,
and Adeeb Shehzad

Abstract

Artificial intelligence (AI) has become a disruptive technology within modern pharmaceutical research and fundamentally overhauls the pharmaceutical discovery, development, and regulatory science paradigms. This chapter presents a comprehensive study of the scientific, operational, legal, and ethical implications of the implementation of AI in the entire drug-level process. During the early discovery stage, AI approaches, such as deep learning, generative modeling, graph neural networks, and virtual screening are improving target discovery, enhancing transition of hits to leads, and facilitating rational design of molecules on a scale never seen before. These methods improve prediction accuracy of physicochemical and ADMET properties, minimize reliance on laboratory tests (which are subsequently costly and labor intensive), and lower overall failure rates that typify preclinical pipelines in the past. With advancing maturity into clinical setting, AI assists in optimization of protocols, stratification of patients, adaptive trial designs, and detection of safety signals by real-time analytics based on electronic health records, image modalities, and biomarker data. Besides, AI-based manufacturing systems are enabling the predictive maintenance, real-time quality control, and constant optimization of the processes in terms of Quality-by-Design (QbD) and Good Manufacturing Practice (GMP) models.

T. Baig · M. J. Khan
Department of Biomedical Engineering and Sciences, School of Mechanical and Manufacturing Engineering, National University of Sciences and Technology, Islamabad, Pakistan

A. Shehzad (✉)
Biodiversity Research Center, Dhofar University, Salalah, Oman

Y. Bibi et al. (eds.), *Artificial Intelligence in Phytochemistry*,
https://doi.org/10.1007/978-981-92-0875-3_14

Tools based on AI, especially machine-learning-based models and adaptive algorithm models, raise auditing and compliance issues, particularly in respect to traceability, prejudice, accountability, and conformity to worldwide data protection frameworks, including the GDPR. This chapter also examines the approach embraced by major regulatory authorities, such as the European Union (EU), the U.S. Food and Drug Administration (FDA), and the United Kingdom (UK) and outlines various divergent philosophies, the emerging direction towards risk-based assessment, lifecycle analysis, and Good Machine Learning Practice (GMLP). To implement responsible, trustworthy deployment of AI, it is required that they are heavily validated, have strong data governance infrastructure, are transparent, audited in fairness, and subject to adaptive regulation like sandboxes. Together, these elements are the basis to leverage the maximum potential of AI and preserve the rights of the patients, the scientific integrity, and social confidence in the changing situation in the sphere of drug development.

Keywords

Virtual screening · Target discovery · Hit-to-lead transition · Rational drug design · Physicochemical

14.1 Introduction

Introduction of artificial intelligence (AI) in drug discovery and development has become one of the most significant changes in contemporary pharmaceutical research [1]. In the recent years, with the increasing computational power, big biomedical data, and improvements in machine learning (ML) and deep learning (DL), AI systems have begun to support or even guide the early-stage target discovery, compound screening, lead optimization, and safety prediction steps [2, 3]. These capabilities are expected to make huge time and cost savings over compared to traditional, highly manual drug discovery processes [4]. In preclinical development, AI tools contribute to in silico models that predict absorption, distribution, metabolism, excretion, and toxicity (ADMET) properties of candidate molecules, potentially decreasing reliance on animal testing and accelerating the selection of safer compounds for clinical studies [5, 6]. In parallel, AI-based drug repurposing, particularly the application based on the existing drug and biological data, represents a potential path toward the re-evaluation of approval drugs with novel indications faster than conventional de novo discovery [7].

At the clinical trial phase, AI-based analytics are applied more frequently to protocol design, optimization of patient recruitment, stratification, and safety or efficacy predictions to enhance the success rate of trials and minimize patient dropout [8, 9]. In the post-marketing process, AI has the capability to assist with the pharmacovigilance process through the analysis of real-world data sources, electronic health records (EHRs), literature, and spontaneous adverse event reports to identify early warning signs of safety issues than what conventional methods would do [4, 10]. Although scientific and operational AI application in drug development has great potential, this shift creates significant legal and regulatory issues. Traditional regulatory structures have traditionally been based on linear, human-reviewable working processes, with properly documented statistics and explicit decision-making criteria [9]. Conversely, many AI/ML models act as black boxes, evolve over time, and demand large amounts of data (many are not homogeneous), thus making traceability, explainability, and validation complex [4, 10].

Current complications signify a leading enhancement between miscellaneous and self-governed responsiveness of artificial intelligence and format, prescriptive guidelines formations which are the premises of medication developments. The existing regulatory frameworks were initially designed to control methodologies that are evident, cold, and readily verifiable, but the existence of a large variety of AI and machine learning models are dynamic, probabilistic systems that remain adaptive to newly introduced information. Such a deviation casts substantive doubt on the sufficiency of current regulatory frameworks, especially in terms of model validation, explainability, data management as well as responsibility when it comes to AI-assisted decision-making. The need to closely reconsider the effectiveness with which existing legal and regulatory frameworks can adapt to these technological changes, without weakening patient safety, scientific soundness, or trust, is increasing as the pace of AI implementation picks up. This chapter is thus aimed at taking a systemic look at such emerging issues, evaluate their ramifications throughout the drug development continuum, and clarify evidence-based measures the integration of AI into pharmaceutical research and development should follow to ensure responsible and transparent and ethically sound.

14.2 AI in the Drug Development Lifecycle

The pillars of artificial intelligence, such as data, computing power, and algorithms, all contribute to leading to the modern use of AI in the field of pharmaceutical research and development. In the drug development ecosystem, the data can come through a very large range of sources such as publicly available biomedical repositories, commercially available datasets, privately generated research and development (R&D) data through industry partnerships, and internally generated experimental data [11]. Other inputs may source up to the scale of entry-data mining programs and then subject the data to intensive data curation, purification, and validation procedures to make it amenable to modeling. The fast advances in computation infrastructure, especially the emergence of high-performance cloud computing and graphics processing unit (GPU)-based systems has also made AI systems capable of scale-based processing of complex biological and chemical data [11]. On the algorithmic level, varying modeling styles—machine learning classifiers, deep learning architectures, and generative models—are chosen and tailored to the scientific question and stage of drug development they will be useful in. The interaction of specialized algorithms and unique data holds will eventually determine the competitive and functional identity of AI-based pharmaceutical technology companies [12].

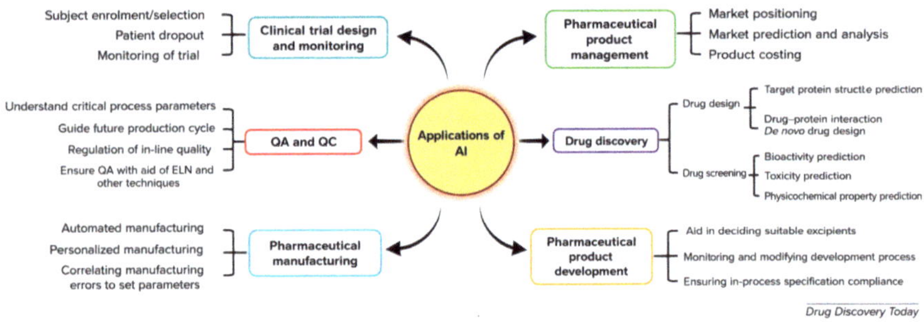

Fig. 14.1 Applications of AI in drug development. (Adapted from Ref. [1])

The figure outlines the broad uses of artificial intelligence (AI) throughout the full lifecycle of pharmaceutical products. AI is used in drug discovery to accelerate many processes, such as identification of therapeutic targets, prediction of protein-ligand interactions, generation of de novo molecular structures, prediction of toxicity, and estimation of physicochemical properties [1]. In the process of pharmaceutical product development, AI is used to select the right excipients, optimize the formulation environment, and ensure compliance with in-process standards. AI-based manufacturing systems also allow automated and personalized manufacturing, identify deviations by matching errors with essential parameters, and improve the efficiency of the entire process. Artificial intelligence tools that can track key process variables, control quality inline, and guide future manufacturing cycles bring benefits to quality assurance (QA) and quality control (QC) through the help of digital laboratory technologies [1].

In the clinical trial design and monitoring field, AI has been used in participant selection, tracking enrolment, predicting dropouts, and ongoing trial monitoring. Besides, AI supports pharmaceutical products by forecasting market, analysis of competition, positioning of products, and predicting costs. Overall, the figure demonstrates AI being incorporated into all stages of pharmaceutical innovation, thus, strengthening scientific and operational decision-making [1] (Fig. 14.1).

14.2.1 Discovery and Target Identification

Drug discovery has been characterized by a high level of chance, and many of the historic discoveries in therapeutic pasts have been followed in both cases as a result of accidental laboratory findings or accidental outcomes of experiments [13]. The chemical space relevant to pharmaceutical discovery is vast, as it is estimated that there are over 10^{60} potential molecular buildings [14]. This tremendous heterogeneity provides significant chances in the discovery and development of new therapeutic agents. However, without advanced computational technologies, exploring such chemical space is a time-consuming, expensive, and resource-constrained task, which traditionally constrained the productivity of traditional drug development pipelines [15]. The solution lies in artificial intelligence (AI) that aids in enabling high-throughput analysis and intelligent prioritization of promising chemical entities. Pattern recognition through machine learning will allow AI systems to quickly pinpoint hit and lead compounds, which will accelerate biological target validation. In addition, AI-based modeling software contributes to the rational optimization of molecular structure, complementing potency, selectivity, and pharmacokinetic properties of the molecule at a more efficient rate, as compared to any other iterative approach [14, 16]. Artificial intelligence (AI) adoption in early research processes would yield significantly lower levels of such unpredictability through the systematic analysis of large chemical and biological data, producing higher chances of yielding drug candidates with commercial and therapeutic potential, at the same time reducing the cost and development time. In 2022, a forecasting study indicated that tremendous investments in AI technologies would bring a greater than 45 % increase in the return on investment within the pharmaceutical industry [17].

Drug development starts with the identification of disease-relevant molecular targets, moves through the identification of active compounds, and ends with the medical preclinical and clinical testing of optimized lead compounds. The route will be fundamentally lengthy, resource-consuming, and risk prone. Estimates of costs show that the process of taking one new therapeutic to market may cost between 100 million and 2 billion dollars and may take up 10–17 years to develop one drug. The attrition rate remains high despite initial signs of progress; for instance, even after successful passing of phase 1 clinical trials, the likelihood of a candidate drug's eventual regulatory approval and market introduction is around 5% [18]. Historically, quantitative structure activity relationship (QSAR) models have been used to rapidly model physicochemical properties (including the partition coefficient (log P)) and distribution coefficient (log D)) and to screen large collections of compounds. However, traditional QSAR methods are often hindered by the inability to model complex biological processes, such as efficacy and safety indices, and limited metrics on small and/or noisy training samples and by lack of experimental predictability [19]. Over the last few years, advanced artificial intelligence (AI) schemes, and especially deep learning (DL), have emerged as strong competencies to measure drug-molecule safety and efficacy through massive data modeling. One bright example is a 2012 Merck QSAR machine learning challenge, which demonstrated that DL models significantly outperformed traditional machine learning algorithms over 15 absorption, distribution, metabolism, excretion, and toxicity (ADMET) datasets, and thus can support future research in developing drug discovery processes at an early stage and prioritization of drug candidate [20].

More effective profiling is made achievable using various computational approaches, including virtual screening of chemical libraries as well as ligand-based and structure-based modeling, which enable faster elimination of non-lead candidates and cost-effective selection of promising drug molecules [14]. Drug design algorithms like Coulomb-matrix representations and molecular fingerprinting have addition of physicochemical and toxicological characteristics to allow identification of leads into development [21]. Computational parameters, including predictive modeling accuracy, molecular similarity, generative design approaches, and other in silico procedures, can convey the predictability of optimal chemical structures to novel compounds [22]. As an example, Pereira et al. presented DeepVS that demonstrated strong performance in docking with 40 receptors and 2950 ligands, and retained good accuracy when pitted against 95,000 decoy structures [23]. Another study used a multi-objective automated substitution strategy to optimize a cyclin-dependent kinase-2 inhibitor, considering shape similarity, biochemical potency, and physicochemical properties [24].

14.2.2 Molecular Characterization Learning

The traditional methods of molecular characterization generate huge numbers of descriptors; nonetheless, they suffer because they depend on user-selected features and are limited by the deficiencies of algorithmic structures. This makes it difficult to determine which structural properties are best suited to support downstream drug development goals [25]. This hinders the effective prioritization of chemical molecules in early stages of discovery [26]. Graph neural networks (GNNs) minimize these difficulties by encoding molecules through graphs, where atoms are nodes and bonds are edges, thus, chemical structures can be derived entirely in the form of matrices or arrays [27].

Every single atomic or bond object is given a feature vector that captures salient chemical properties, such as the atomic species, formal charge, the number of hydrogen atoms, bond order, and ring membership [28]. This kind of automated learning of features is particularly important in pharmaceutical development where tiny structural changes often control the levels of potency, selectivity, and toxicity. Graph neural networks can learn molecular representations by using graph convolution without relying on handcrafted descriptors, thus showing strong results in the prediction of physicochemical and biological properties [29]. Neural FP, the Weave architecture, and Attentive FP algorithms can capture long- and short-range interactive correlations to improve predictions of solubility, bioactivity, and toxicity-related parameters, which are especially significant in the initial selection of drug candidates [30, 31]. In other applications of property prediction, graph neural networks play a vital role in ab initio molecular design, in modeling drug-target interactions, in inverse drug-discovery workflows and allow researchers to find and optimize promising chemical entities more efficiently [32].

14.2.3 Prediction of Physicochemical Properties

The essential physicochemical characteristics, specifically, solubility, partition coefficient (log P), ionization state, and intrinsic permeability, have an enormous impact on the pharmacokinetic profile of a drug and on the interactions of the compound with receptor families, thus, being the inseparable factors to consider during the initial phases of drug design [33]. The forecasting of such properties is now possible through artificial-intelligence-based tools that internalize the knowledge of extensive historical compound-optimization data sets [34]. Modern drug-design algorithms use numerous molecular descriptors such as SMILES representations, potential-energy profiles, electron-density measurements, and 3D atomic coordinates, to produce chemically feasible structures using deep neural networks (DNNs), and subsequently the properties of the resulting structures are approximated [35].

A workflow has been developed by Zang et al., which is named a quantitative structure-property relationship (QSPR) and predicts six physicochemical parameters of environmental chemicals through the U.S. Environmental Protection Agency Estimation program Interface (EPI) Suite [34]. Applications of neural networks to ADMET Predictor and ALGOPS have also been used to predict lipophilicity and solubility in a significant subset of chemical classes [36]. Graph-convolutional neural networks (GCNN) and undirected graph recurrent neural networks are additional types of deep learning frameworks that have been utilized in order to enhance the accuracy of solubility predictions [37]. Graph-kernel methods, auction neural network (ANN) models, and kernel hinge-ridge regression frameworks have been used to predict acid dissociation constants (pKa values) of a wide range of compounds [34, 38]. Cellular permeability in model systems, such as Madin Darby canine kidney (MDCK) and Caco2 cell lines, could be experimentally measured, which would serve as input into AI-assisted predictive models of membrane transport activity [39]. All of these advances highlight the growing use of artificial intelligence in predicting physicochemical and related bioactivity profiles in drug design and development.

14.2.4 Non-clinical Development and Safety Assessment

The next critical phase after the identification of candidate molecules is the non-clinical testing (the assessment of absorption, distribution, metabolism, excretion, and toxicity (ADMET)) and off-target risks and other safety variables [40]. Traditional toxicological and pharmacokinetic assays are labor intensive, lengthy, and are often ethically questionable (especially animal testing). The potential of artificial intelligence is particularly high in this area: as recent systematic reviews show, contemporary machine learning and deep learning models can outperform the traditional quantitative structure-activity relationship (QSAR) framework on predicting organ-specific toxicities, including hepatotoxicity and cardiotoxicity, genotoxicity, and other undesirable events, through training on large, high-dimensional chemical and biological datasets [1, 40].

Moreover, AI supports the combination of multiparametric data, such as molecular structural descriptors, in vitro assay outcomes, and transcriptomic, proteomic, and metabolomic data, resulting in the overall toxicity risk profiles, which can possibly predict the potential latent or rare adverse effects more sensitively than the traditional models [41]. Such multimodal schemes have the potential to reduce the rates of attrition in the initial developmental phases by filtering out high-risk candidates earlier before initiating the costly in vitro or in vivo research [42].

Such developments promise considerable potential to combine safety and efficiency, but at the same time, they also present relevant regulatory and legal issues.

Regulatory bodies, based on the principles of Good Laboratory Practice (GLP) and Good Manufacturing Practice (GMP), must be able to determine whether the AI-generated in silico evidence can be considered as acceptable data to be included in regulatory submissions or it is simply supportive, non-definitive analysis [40]. Such evidence is conditional in that it depends on hard validation, which includes clear documentation of the model, data provision, performance measures, and uncertainty quantification and suitability to the purpose of the overall practice, as does first-in-human dose selection or toxicity prediction [41].

The other relevant issue is the management of model lifecycles. In contrast to the deterministic experimental framework, machine learning–trained model is continuously refined with new input data and retrained. Such adaptive dynamics bring additional challenging problems in the form of reproducibility and traceability that are crucial to the need to comply with the regulations. Use of adaptive AI systems in safety assessment procedures can only be reviewed to meet regulatory expectations with strong version control, extensive audit trail, systemic revalidation of each update, and careful documentation of provenance of information and model updates [41]. The interpretation of legal liability is still uncertain: at the moment when the AI-generated safety profile turns out to be inadequate and the compound passes to human testing, it is unclear whether the sponsor, the AI developer, or both are liable. Through this uncertainty, some researchers have proposed that purpose-designed regulatory-scientific systems of surveillance, indemnities, and contractual arrangements should be implemented to curb such risks [41].

Overall, the potential of AI-based non-clinical safety evaluation is huge in terms of efficiency promotion, animal testing minimization, and early risk identification, but its implementation depends on its incorporation into a regulatory framework, which demands strict validation, clear documentation, effective governance processes, and unquestionable liability lines.

14.2.5 Clinical Trial Design, Conduct, and Analysis

When candidate selection is progressed to human testing, artificial intelligence (AI) is more and more being utilized to supplement every area in the clinical trial design, trial process, and clinical trial analysis. Machine learning (ML) and deep learning algorithms can be used to design optimal protocols, recruit patients, randomize on the fly, dynamically adjust trials based on interim results, and even change protocols based on real-time monitoring of safety and efficacy signals. One of the latest reviews has outlined the way AI-driven trial-matching algorithms identifying and processing electronic health records (EHRs) significantly decrease recruitment time and increase enrolment accuracy, decreasing both screen-failure rates and shortening recruitment timelines [10].

In addition to recruitment, AI-aided adaptive trial designs use predictive models and Bayesian modeling to stratify the patients, predict the risk or treatment response of individual patients, and dynamically adjust allocation which might enhance statistical power, decrease sample sizes, and enhance safety monitoring. Imaging or biomarker-based endpoints controlled by AI can be used in some settings to detect efficacy or toxicity earlier, therefore reducing the trial time and increasing the sensitivity [43]. These potent power features come at a huge regulatory and moral cost. In cases where AI tools influence eligibility, randomization, monitoring, or endpoint adjudication, they are essentially part of what is referred to as the so-called computerized system under the rubric of Good Clinical Practice (GCP). Such systems are now recommended by regulators such as the European medicines agency (EMA) to be validated, have audit trail, have controls in data integrity, human oversight and documentation that is transparentized and commensurate to their degree of control [44].

The major issue is bias and fairness. Trained artificial intelligence models will often be under-representative of certain demographic or socio-economic groups, causing recruitment bias, unequal representation, or an incorrect signal of safety and efficacy not representative of the overall population [45]. Without the direct demographic performance audit and sound bias mitigation strategies, the generalizability and equity of clinical trials will be at risk. Various academic researchers have emphasized the necessity to embrace demographic transparency, conduct the audit of fairness, and stipulate adaptive courses that will amalgamate the under-represented groups in AI-directed recruitment and stratification [45].

Ethical requirements also entail transparency with participants. Using AI to screen, enroll individuals, monitor, or engage in adaptive decision-making should be clearly revealed in informed-consent legality societal forms. These disclosures ought to clarify that inclusion, continuation, or disengagement during the trial may be subject to algorithmic choices, and must provide future opportunities to review them by humans. The lack of such transparency can undermine the integrity of participants and damage the public trust [45].

Security, privacy, and data integrity are the salient issues when implementing artificial intelligence in clinical research. Clinical trial data are generally the most confidential records of individual health, and AI systems that handle such data must comply with data protection laws rigorously, utilize powerful encryption, retain elaborate audit histories, and have strict access-control measures. Weaknesses in governance increase the risk of data breaches, non-compliance with regulations, and the loss of patient trust. Therefore, due to its transformative capacity to improve the efficiency of trials, statistical power, and adaptability, the integration of AI into clinical development should necessitate strict validation procedures, robust documentation, shield against bias and inequity, explicit informed consent of subjects, and careful data utilization supported by suitable regulatory controls.

14.2.6 Manufacturing and Quality Control

In addition to the discovery and clinical stages, AI also has implications in the pharmaceutical production and quality control. The applications of AI and machine learning algorithms to optimize process parameters, inspection during the production line, defect inspection, batch failure prediction, and real-time quality control are becoming more popular in modern manufacturing environments. This includes continuous manufacturing, high-throughput production lines, and complex formulation processes under Quality-by-Design (QbD) paradigms [46]. Indicatively, AI-based chemometric training and sensor-data analysis can detect flaws in pills (cracks, discoloration, cracks) and forecast deviations in uniformity as well as design abnormalities in formulations significantly earlier in advance compared to conventional manual testing or end-of-the end product analysis methods [42, 46].

Moreover, artificial intelligence–based predictive maintenance and predictive process monitoring helps decrease variability, increase reproducibility, and optimize materials and process throughput, which also helps to achieve more reliable and consistent product quality, a decrease in rejects, and lower the cost of production [47]. However, the integration of AI in controlled production and quality systems poses serious challenges in regulation. These types of AI/ML systems, which affect the quality or release of the batch in the pharmaceutical industry, should be incorporated into the Pharmaceutical Quality System (PQS). As such, these types of models must be thoroughly validated with specifications intended use, performance measurement, robustness testing, version management, change-control measure, audit trails, data integrity assurances, and documented risk evaluation [41, 42].

Artificial intelligence (AI) models that are adaptive or constantly learning are very difficult to regulate. Such models could improve performance as the production data increases, but because of its inherently dynamic nature, they are incompatible with regulatory demands of stability, reproducibility, and fixed validated processes, which are cornerstones of Good Manufacturing Practice (GMP) [48]. Without strict change-control mechanisms and re-validation procedures whenever a model is changed, the manufacturing process becomes unpredictable thus undermining the integrity of the batches and compliance with legal regulations [49].

Although AI-supported production and quality management may bring significant positive changes in efficiency, consistency, and overall costs, their regulation acceptance will depend on their implementation into the existing Product Quality System (PQS) frameworks. Strict validation, resilient governance, and proximate human monitoring especially in cases where models are made to affect key quality or safety features are essential to the attainment and sustenance of regulatory approval.

14.2.7 Post-authorization and Pharmacovigilance

Safety surveillance, or pharmacovigilance (PV), is still necessary following a pharmaceutical agent approval and subsequent availability on the market. The voluminous, heterogeneous, and complex character of real-world data, such as spontaneous adverse event reports, electronic health records, registries, scientific literature, and even social media content, is a tremendous challenge to traditional PV systems. Machine learning (ML) and artificial intelligence (AI) have the potential to provide effective answers to these problems [50]. Nagar and colleagues review of the adoption of AI in the pharmacovigilance field, published in 2025, determined that

AI-based systems are able to significantly enhance the speed, sensitivity, and performance of adverse-event detection and hence identify safety signals earlier than traditional disproportionality-based systems [51]. In practice, AI-based pharmacovigilance systems have been shown to process both structured and unstructured data: they are able to automatically code adverse events, duplicate reports, isolate relevant clinical information in narrative text, such as clinical notes, discharge notes, social media text, and literature, and flag serious or unexpected events to human review—reducing the manual workload and improving the level of detection sensitivity [52].

Machine learning (ML) and knowledge graph (KG) paradigms can offer a framework of the mechanistic hypothesis generation. This can be done through a combination of drug chemical structures, known target features, human-metabolic-pathway data, and adverse-reaction database [53]. Artificial intelligence (AI) systems would be able to suggest biologically plausible mechanisms potentially explaining observed toxicities or drug-drug interactions, improving regulatory risk assessment procedures, and assisting post-marketing investigations. Transparency and regulatory trust are further enhanced by explainable AI (XAI) processes that provide interpretable reasoning for the associations flagged by the model [54].

However, there are a number of regulatory, ethical, and operational concerns. Regulatory language emphasizes that AI-based pharmacovigilance needs to be used as an addition and not a substitute to human causality judgment and judgment by experts. Even in cases where an AI-driven system raises a red flag, as a possible warning of apparent safety, the end-calls involving risk alleviation, labeling changes, or market resources should be made across the board of human capabilities, clinical scrutiny as well as regulatory guidelines [44]. Overall, the concept of the artificial intelligence technologies integration into the sphere of pharmacovigilance has a strong potential to promote the post-market safety surveillance, reduce the number of adverse events identified earlier, and support the adaptive risk management approach. However, these advantages depend on the adoption of sound governance systems that will ensure human control, transparency, information integrity, version management, and regulatory responsibility.

14.3 Global Regulatory Landscape for AI in Drug Development

Within the framework of global jurisdictions, the use of AI in healthcare is acknowledged by existing medical device laws in the most, with particular reference to the regulatory framework of the Software as a Medical Device (SaMD) [55]. Furthermore, one should take into consideration that these frameworks do not include specific types of AI tools, such as applications aimed at promoting healthy lifestyles, systems that only cater to administrative tasks, and software, which provides clinical support or advice to medical staff. This exclusion has been justified by the assumption that such contexts allow users to use the requisite professional competence or personal autonomy to interpret AI-generated recommendations and to make informed and autonomous choices [56].

The regulatory landscape for AI in healthcare differs widely across nations, shaped by diverse priorities that include ethical principles, security considerations, data privacy protections, economic strategies, and cultural norms [57]. While China is also a major player in the global AI ecosystem, the United States (USA), the European Union (EU), and the United Kingdom (UK) have emerged as the leading jurisdictions in developing regulatory frameworks that influence international standards for these technologies [58].

14.3.1 European Union, Europe

AI regulatory efforts within the EU are primarily designed to establish a secure, trustworthy, and accountable environment for the development and deployment of AI systems [59]. The EU has introduced one of the most extensive and structured regulatory frameworks for AI in the healthcare sector, grounded in the General Data Protection Regulation (GDPR) and complemented by the proposed AI Act, which came into force in 2018. Under this framework, GDPR obligates healthcare institutions and AI developers to obtain informed consent before processing patients' data and to implement robust safeguards that ensure data minimization, purpose limitation, and overall data security.

A comprehensive framework of regulation was proposed in April 2021 by the European Commission under the name AI Act that should govern the security, ethical, and human rights issues in all EU member states. The Act uses a risk-based paradigm and identifies four levels of risk in AI systems, including unacceptable, high, low, and minimal, with the majority of the AI applications applied to healthcare falling into the high-risk category based on their risks to clinical decisions, patient safety, and overall healthcare outcomes. High-risk AI systems under the AI Act should meet rigorous standards of data governance, of risk management, and of continued oversight. It is a duty of developers and manufacturers to act with integrity, transparency, accountability, and responsible behavior in the lifecycle of the system [59, 60]. Moreover, the Act directly prohibits AI use cases that imply unreasonable risks, including instances that can jeopardize basic human rights or safety. It further offers a systematic methodology to the classification, regulation, and liability sources of high-risk AI systems, such as those employed in healthcare, and structures regarding general-purpose AI (GPAI) models [61].

With the help of the Medical Devices Regulation (EU) 2017/745 (MDR) and the In Vitro Diagnostic Medical Diving Regulation (EU) 2017/746 (IVDDR), the European Union has adopted a certain regulatory framework on health and medical products, including those that include artificial intelligence. The regulations subject the assessment processes in pre-market procedures with more rigor, clinical evidence with stronger requirements, and post-market surveillance with greater obligations in the full lifecycle of a medical device. The key aim behind such steps is to make sure that AI-powered medical technologies are always up to high healthcare safety, performance, and reliability standards [62].

In the MDR, medical devices are classified into four classes that describe the degree of possible risk to users: Class I, Class IIa, Class IIb, and Class III, which is low risk, medium risk, high risk, and high risk, respectively. Class III devices are those that result in the strictest regulatory control that includes extensive conformity, vigorous clinical trials, and unrelenting supervision to ensure continued adherence to both safety and performance standards [60, 62].

14.3.1.1 EMA

The European Medicines Agency (EMA) has significantly preempted the contribution of artificial intelligence (AI) and machine learning (ML) to its regulatory science agenda [63]. To expand this effort, the Agency has established a series of cross-agency coordination teams focused on AI, whose goals are to enhance the oversight of AI technology, enhance harmonization between departments, map internal capabilities, and define mandates in AI-related actions. Under these larger programs, current discussions within the EMA AI technical groups revolve around increasing knowledge sharing, researching in-house, and generating recommendations and best-practice advice to accelerate innovation in data science and AI [64].

One of the most critical engines of digital innovation at EMA is the Digital Business Transformation Task Force, which serves as a central repository in the experimentation, design, and implementation of digital transformation initiatives. This taskforce oversees the whole process of digital adoption, including conceptual strategy and initial design, pilot test, and ultimate rollout. EMA also provides a range of AI-assisted services, as part of this program, such as process re-engineering and digital process mapping, to streamline workflows and use automation and AI technologies to enhance operations [63]. The initiative is active in surveying emerging digital technologies, appraising proof-of-concept opportunities, and finding technologies with the capabilities to produce quantifiable organizational gains, including improved efficiency.

Parallel to this, the Analytics Center of Excellence (ACE) is located at EMA and focuses on researching how advanced analytics can be used in practical solutions to the regulatory and operational needs of the Agency; this includes AI, ML, and robotic process automation [64]. The overall objective of ACE is to implement data-driven instruments to improve performance and simplify a regulatory procedure. In addition, EMA is investing heavily in its healthcare data analytics capacity. One of the priority areas in this technology is how AI, ML, and new digital technologies can organize unrefined or complex data and provide more profound and actionable insights into healthcare data [65].

14.3.1.2 Food and Drug Administration (FDA), USA

In the United States, there is a decentralized paradigm of artificial intelligence regulation in which various federal organizations can play a role in supervisory roles. However, the leading authority in the regulation of AI-driven technologies in healthcare is the Food and Drug Administration (FDA). In partnership with the

Center of Biologics Evaluation and Research (CBER) and the Center of Drug Evaluation and Research (CDER), the FDA has been developing regulatory approaches that are explicitly tailored to AI-driven medical devices and Software as a Medical Device (SaMD) used in the clinical environment. FDA has an overall strategy that is based on a risk-predictable classification framework, as the amount of regulatory attention is proportionate to the risk potential that an AI system poses to patient safety [66].

In April 2019, the Food and Drug Administration published its "Proposed Regulatory Framework Changes to AI/ML-Based SaMD" which included a model based on the total product lifecycle (TPLC). This model recognizes that AI/ML systems are not static but dynamic in nature and can evolve during the exposure to new data sets. Developers are therefore tasked with ensuring the long-term realistic operation of their algorithms; furthermore, any change that changes the purpose with which the system is supposed to perform requires a fresh cycle of engagement back to the path of approval [67].

Continuing its regulatory paradigm, the FDA presented the Digital Health Software Pre-Certification (Pre-Cert) Program in 2021 [67]. This initiative will focus on developer testing and certification, unlike traditional product-based assessments, which are developer-focused. There is a possibility that organizations with strong organizational excellence, strict quality assurance measures, and evident dedication to patient safety can enjoy a fast pathway to market entrance, but they will still be under constant scrutiny and performance control [68].

14.3.1.3 Regulatory Authorities, United Kingdom

In 2019, the National Institute for Health and Care Excellence (NICE) published the Evidence Standards Framework of the Digital Health Technologies, in cooperation with NHS England. The framework provides guidelines on how to evaluate a broad range of digital health solutions, such as mobile health applications, clinical decision-support software, and web-based platforms, whether such technologies are implemented independently or as part of other medical products [69].

In November 2022, the Regulatory Horizons Council, an advisory body providing strategic advice to the government of the United Kingdom on new technological trends, released its report titled The Regulation of AI as a Medical Device [70]. The document takes a lifecycle-based approach on AI-enabled medical devices (AI-MDs). One of the key focal points of this guidance is enhancing the presence of patients and the general population in the oversight mechanism, therefore, bringing about more transparent and efficient communication among regulators, industry creators, and end users [70].

Similarly, the Medicines and Healthcare Products Regulatory Agency (MHRA) introduced the Software and Artificial Intelligence as Medical Device Change Program in September 2021. The program is proposed to redefine and strengthen the laws that regulate AI-based medical technologies in the United Kingdom by establishing a clear regulatory path. It is organized into two large workstreams. The initial workstream concerns reforms throughout the lifecycle of Software as a Medical Device (SaMD) with specific emphasis, though not limited to, on cybersecurity threats, data protection requirements, and additional post-market oversight regulation. The second workstream focuses on the specifics of AI systems, especially the flexibility of machine learning algorithms, the dangers of the algorithmic inequity, and the necessity of improved interpretability and transparency of AI models [70].

14.3.2 Global Regulatory Convergence Challenges

The harmonization of artificial intelligence regulation in the healthcare field is a global but intricate and fragmented process. The major barrier is the presence of substantive divergent privacy laws, in particular the difference between the comprehensive General Data Protection Regulation (GDPR) of the European Union and the more sector-focused, decentralized regulatory paradigm of the United States. These conflicting legal systems create significant issues in multinational development and data ingestion involving AI systems. These obstacles are, in essence, due to opposing regulatory ideologies: the European Union is more focused on human rights and precautionary ideals, and the United States is focused on innovation and more relaxed attitude toward risk. Furthermore, other agencies, like the World Health Organization, take a more comprehensive approach to equity and ethical concern [71].

In low- and middle-income countries (LMICs), there are other challenges concerning the heterogeneous distribution of AI regulation frameworks and are typically either immature, or unevenly distributed, which leads to a heterogeneous ability to observe such structures. International limits on cross-border data transfers also hinder the joint development of AI since they limit the export of sensitive health data needed to effectively train models [72].

Another gap is the lack of generally recognized standards governing the documentation and validation of the AI systems, which is comparable to the properly developed principles like CONSORT or SPIRIT used in clinical research. A combination of these disparities prevents coordinated global regulation and makes assessment, implementation, and surveillance of AI application in healthcare more complex [65].

14.4 Data Protection, Privacy, and Data Governance

14.4.1 Nature of Data Used by AI in Drug Development

AI systems utilized in drug discovery and development are typically based on a heterogeneous combination of real-world data (RWD), electronic health records (EHRs), claims data and trial datasets, along with omics and imaging data. Chen et al. describe the gradual adoption of AI across the drug development lifecycle and the importance of RWD and EHRs in identifying targets, trial optimization, and pharmacovigilance, using the example of several, and numerous, therapeutic fields [73].

The recent reviews underscore the fact that AI-based drug development also involves the use of highly sensitive genomic, transcriptomic, and proteomic profiles, and digital pathology images, all of which present a high risk of re-identification when they are combined with clinical history and demographic factors [74].

This rich multimodal data is often longitudinal in the context of precision medicine and personalized drug programs, which makes it possible to model disease head processes fine-granularly, but it also increases the risk of individual single-person identification, especially in rare diseases or small studies [18].

14.4.2 Lawful Bases for Processing and Secondary Use

A lawful processing of health and genetic data to develop AI following the GDPR should have a lawful basis under Article 6 and should meet the stricter safeguards applicable to special categories of personal data under Article 9; normally by explicit consent, the performance of activities in the public interest (such as scientific research), or, in some circumstances, by legitimate interests and very strong safeguards (Regulation (EU) 2016/679) [75]. The European policy tools on the reuse of health data as secondary use now fall under the umbrella of the proposed European Health Data Space (EHDS) with the reuse of health data for research and innovation allowed as long as principles of purpose limitation, proportionality, and respect of data-subject rights are followed [75].

The legal scholarship has discussed the most suitable legal grounds to use in the secondary purposes like AI model training. Becker et al. believe that scientific research involving health and genetic data frequently depends on consent and public interest, cross-border projects need more clarity on how these frameworks apply to them [76]. The interaction of GDPR and Data Governance Act with future EHDS regulation in the area of secondary use is discussed by Shabani et al., where the authors pay attention to the necessity of the appropriate governance framework and clear legal foundations in the projects involving AI [77].

The broad consent is central in the management of longitudinal cohort studies and biobanks. Zenker et al. map out the recent German national broad-consent model, which is designed to be in line with the General Data Protection Regulation but at the same time allow the secondary use of the healthcare-related data and bio-specimens in biomedical research, including artificial intelligence studies [78]. Empirical studies have recently shown that broad-consent templates often do not clarify to potential donors about all the ways their data can be used, potentially undermining confidence in AI-based re-use of healthcare data [79, 80].

14.5 Governance, Equity, and Future Directions for AI in Drug Development

Recent research indicates that AI used in drug discovery and drug development should be controlled as a part of the pharmaceutical quality system and not a separate information technology experiment. Analysis of international AI regulation in the healthcare sector reveals that there has been a sharp transition to risk-based classification of AI applications, proportional validation of computerized systems, and integration of AI governance into current GxP models [71].

This means that sponsors must identify, e.g., between low-impact internal decision-support models and high stakes tools which may affect the protocol design, eligibility decisions, or safety observations. Comparative law reviews of FDA and EMA regulation also argue that any significant use of AI regulation requires denoting clear splitting of roles between R&D, clinical teams, biostatisticians, regulatory affairs, quality assurance, and data protection officers with collective responsibility of documenting the model, performance monitoring, and responding to incidents across the lifecycle [81]. Viewed through the prism of an integrated governance framework, the principles similar to the Good Machine Learning Practice (GMLP) are increasingly being applied not to medical devices but to artificial intelligence systems used in clinical trials and pharmacometric experiments. Thematic areas that keep reoccurring in modern systematic reviews include a clear definition of the targeted and situational use, the reliance on representative and quality training data, comprehensive internal and external validation based on appropriate performance indicators, and a keen interest in usability and the interaction between humans and artificial intelligence in real clinical practice [82, 83].

Risk management regulatory science writers theorize a lifecycle-based approach to risk management. During the planning phase, the sponsors should be able to state the clinical or operational issue, outline data sources, define performance thresholds, and identify ethical limitations. Formidable data management, bias analysis, and resistant data pipelines need to be followed during the development. It should be validated based on preset acceptance criteria, sensitivity analysis, and independent data set. The deployment should be backed by systematic training, well-documented procedures, and user support. Finally, performance drift, fairness indicators, and adverse events should also be continually measured by monitoring, and model iterations should be controlled by formal change-control procedures [71, 83].

Simultaneously, equity and bias have become a key issue in the field of artificial intelligence as a tool in drug development and clinical research. Empirical and conceptual research on fairness has shown that artificial intelligence trained on biased datasets or on geographically limited datasets tends to under-represent certain categories, thus potentially contributing to existing inequalities in diagnosis, treatment, and access to trials [84]. More recent reviews provide the list of the possible points at which the bias can enter the pipeline, such as data collection, labeling, feature engineering, model design, and deployment, and suggest mitigation measures, including using more diverse datasets, fairness-aware learning algorithms, and explicitly tracking the performance across the demographic subgroups [85].

At the same time, equity and bias have taken on a central role in the field of artificial intelligence, acting as leverage in the context of drug development and clinical research. Empirical and conceptual studies have shown that artificial intelligence models that are trained using biased or geographically limited datasets have worse representativity of under-represented groups and therefore may contribute to the existing factors of disparity in diagnosis, treatment, and access to trials [85]. The recent systematic reviews list the possible weaknesses of the artificial intelligence pipeline, such as data collection, labeling, feature engineering, and model architecture, as well as mitigation measures, such as using more diverse datasets, using fairness-conscious algorithms, and expressly staying on track of performance across demographic subgroups [86].

Resting on AI-assisted clinical trials, researchers have warned against algorithms used to conduct site selection and recruit patients having biased results as they favor well-funded centers or patients with extensive digital history at the cost of representativeness and external validity [87]. In turn, it is highly recommended that sponsors consider equity issues in the directions of protocol design and statistical analysis, such as setting diversity goals in advance, being mindful of demographic performance metrics, and capturing the ways in which the outputs of algorithms are used and might be overridden by clinicians in practice [86].

Future research by legal and regulatory theorists indicates that unification around widely held values is possible and even preferable, despite the continued existence of non-homogenous national systems. Comparative studies find that an emergent consensus exists on risk-based regulation, human-centric regulation, strong data governance, transparency, and dynamic lifecycle monitoring, which serve as pillars to support trustworthy AI in the sphere of medicines research and development [81, 88]. However, there are still huge gaps. Cross-regional surveys identify conflict between restrictive privacy frameworks and the volume of information demanded by artificial intelligence, unanswered questions that touch on whose responsibility it is in the intricate watchful AI supply chains, and an apparent absence of uniform reporting forms that could compare AI-facilitated drug development protocols with either the CONSORT or SPIRIT [89].

To counter this, several authors promote AI regulatory sandboxes and other experimental processes in health. These platforms can help sponsors and regulators to co-design assessment guidelines, explore actual performance and equity, and enhance a meaningfully improvement in governance instruments without impeding innovation [90]. Together, these strands encompassing quality-system governance, equity-focus model-development and experimentally grounded, but principled regulatory innovation combine to define a prospective AI agenda in drug development aimed at maximizing clinical benefit and protecting the rights of patients and promoting from the promotion of trust in the community [91].

14.6 Conclusion

Artificial intelligence in drug discovery and development is a paradigm shift in the history of pharmaceutical innovation. In all levels, including molecular design and ADMET prediction, optimization in clinical trials, manufacturing oversight, and post-marketing success, AI provides powerful controls that can speed operations, enhance knowledge, and provide more careful, responsive therapy. However, this chapter emphasizes that such scientific and operational advantages are mitigated by the resultant regulatory, ethical, and legal burdens.

The current regulatory landscape is a major paradox in the entire world full of regulatory agencies, the EMA, FDA, or the MHRA more and more acknowledge that AI systems are dynamic and adaptive, and the existing regulations are stuck in a rut of assumptions based on transparency, reproducibility, and the stability of processes. Novel governance accordingly requires AI-enhanced technologies, in which lifecycle management, rigorous verification, bias reduction, and strong data security are priorities, as well as accountability in its strictest meaning.

The future of artificial intelligence in drug development depends on coordinating regulatory frameworks, adopting open submission procedures, adhering to strong multi-national data, and instituting total governance systems mediated through sympathetic partnership among regulatory bodies, technology researchers, clinicians, and ethicists. Implementation of the Good Machine Learning Practices (GMLP), sandboxes, and embedded pharmaceutical quality systems has a great potential to reconcile innovation and patient safety. Finally, an informational architecture that is human-centered and pragmatic is necessary to support the responsible use of AI; an architectural model must secure the inherent principles of scientific progress and, at the same time, protect the population, ensure data integrity, and provide equitable health services.

References

1. Paul, D., et al. (2021). Artificial intelligence in drug discovery and development. *Drug Discovery Today, 26*(1), 80–93.
2. Blanco-González, A., et al. (2023). The role of AI in drug discovery: Challenges, opportunities, and strategies. *Pharmaceuticals (Basel), 16*(6).
3. Sumathi, S., et al. (2023). A review on deep learning-driven drug discovery: Strategies, tools and applications. *Current Pharmaceutical Design, 29*(13), 1013–1025.
4. Bassey, G. E., et al. (2025). Transformative role of artificial intelligence in drug discovery and translational medicine: Innovations, challenges, and future prospects. *Drug Design, Development and Therapy, 19*, 7493–7502.
5. Kim, H., et al. (2020). Artificial intelligence in drug discovery: A comprehensive review of data-driven and machine learning approaches. *Biotechnology and Bioprocess Engineering, 25*(6), 895–930.
6. Serrano, D. R., et al. (2024). Artificial Intelligence (AI) applications in drug discovery and drug delivery: Revolutionizing personalized medicine. *Pharmaceutics, 16*(10).
7. Rehman, A. U., et al. (2025). Role of artificial intelligence in revolutionizing drug discovery. *Fundamental Research, 5*(3), 1273–1287.

8. Fu, C., & Chen, Q. (2025). The future of pharmaceuticals: Artificial intelligence in drug discovery and development. *Journal of Pharmaceutical Analysis, 15*(8), 101248.
9. Ocana, A., et al. (2025). Integrating artificial intelligence in drug discovery and early drug development: A transformative approach. *Biomarker Research, 13*(1), 45.
10. Kant, S., Deepika, & Roy, S. (2025). Artificial intelligence in drug discovery and development: Transforming challenges into opportunities. *Discover Pharmaceutical Sciences, 1*(1), 7.
11. Antonissen, N., et al. (2025). Artificial intelligence in radiology: 173 commercially available products and their scientific evidence. *European Radiology*, 1–11.
12. Javed, A. R., et al. (2023). Toward explainable AI-empowered cognitive health assessment. *Frontiers in Public Health, 11*, 1024195.
13. Bai, F., Li, S., & Li, H. (2024). AI enhances drug discovery and development. *National Science Review, 11*(3), nwad303.
14. Mak, K.-K., & Pichika, M. R. (2019). Artificial intelligence in drug development: Present status and future prospects. *Drug Discovery Today, 24*(3), 773–780.
15. Mishra, V. (2018). Artificial intelligence: The beginning of a new era in pharmacy profession. *Asian Journal of Pharmaceutics (AJP), 12*(02).
16. Sellwood, M. A., et al. (2018). *Artificial intelligence in drug discovery* (pp. 2025–2028). Taylor & Francis.
17. Belenguer, L. (2022). AI bias: Exploring discriminatory algorithmic decision-making models and the application of possible machine-centric solutions adapted from the pharmaceutical industry. *AI and Ethics, 2*(4), 771–787.
18. Serrano, D. R., et al. (2024). Artificial intelligence (AI) applications in drug discovery and drug delivery: Revolutionizing personalized medicine. *Pharmaceutics, 16*(10), 1328.
19. Zhu, H. (2020). Big data and artificial intelligence modeling for drug discovery. *Annual Review of Pharmacology and Toxicology, 60*(1), 573–589.
20. Ciallella, H. L., & Zhu, H. (2019). Advancing computational toxicology in the big data era by artificial intelligence: Data-driven and mechanism-driven modeling for chemical toxicity. *Chemical Research in Toxicology, 32*(4), 536–547.
21. Chan, H. S., et al. (2019). Advancing drug discovery via artificial intelligence. *Trends in Pharmacological Sciences, 40*(8), 592–604.
22. Brown, N. (2015). *In silico medicinal chemistry: Computational methods to support drug design*. The Royal Society of Chemistry.
23. Pereira, J. C., Caffarena, E. R., & Dos Santos, C. N. (2016). Boosting docking-based virtual screening with deep learning. *Journal of Chemical Information and Modeling, 56*(12), 2495–2506.
24. Firth, N. C., et al. (2015). MOARF, an integrated workflow for multiobjective optimization: implementation, synthesis, and biological evaluation. *Journal of Chemical Information and Modeling, 55*(6), 1169–1180.
25. Simpkin, A., et al. (2024). Using deep-learning predictions reveals a large number of register errors in PDB depositions. *IUCrJ, 11*(Pt 6), 938–950.
26. Torodii, D., et al. (2025). Crystal structure validation of verinurad via proton-detected ultrafast MAS NMR and machine learning. *Faraday Discussions, 255*, 143–158.
27. Mohanraj, P., Raman, V., & Ramanathan, S. (2024). Deep learning for Parkinson's disease diagnosis: A Graph Neural Network (GNN) based classification approach with Graph Wavelet Transform (GWT) using protein–peptide datasets. *Diagnostics, 14*(19), 2181.
28. Duan, P., Zhou, C., & Liu, Y. (2023). Dynamic graph representation learning via coupling-process model. *IEEE Transactions on Neural Networks and Learning Systems, 35*(9), 12383–12395.
29. Fan, Z., et al. (2024). Enhancing predictions of drug solubility through multidimensional structural characterization exploitation. *IEEE Journal of Biomedical and Health Informatics, 29*(3), 1828–1837.
30. Gu, X., et al. (2024). MFD–GDrug: Multimodal feature fusion-based deep learning for GPCR–drug interaction prediction. *Methods, 223*, 75–82.

31. Chen, B., et al. (2024). Is fragment-based graph a better graph-based molecular representation for drug design? A comparison study of graph-based models. *Computers in Biology and Medicine, 169*, 107811.
32. Yuan, Y., et al. (2024). Tripartite interaction representation algorithm for crystal graph neural networks. *Scientific Reports, 14*(1), 24881.
33. Zang, Q., et al. (2017). In silico prediction of physicochemical properties of environmental chemicals using molecular fingerprints and machine learning. *Journal of Chemical Information and Modeling, 57*(1), 36–49.
34. Yang, X., et al. (2019). Concepts of artificial intelligence for computer-assisted drug discovery. *Chemical Reviews, 119*(18), 10520–10594.
35. Hessler, G., & Baringhaus, K.-H. (2018). Artificial intelligence in drug design. *Molecules, 23*(10), 2520.
36. Lusci, A., Pollastri, G., & Baldi, P. (2013). Deep architectures and deep learning in chemoinformatics: The prediction of aqueous solubility for drug-like molecules. *Journal of Chemical Information and Modeling, 53*(7), 1563–1575.
37. Kumar, R., et al. (2017). Prediction of human intestinal absorption of compounds using artificial intelligence techniques. *Current Drug Discovery Technologies, 14*(4), 244–254.
38. Rupp, M., Körner, R., & Tetko, I. V. (2010). Estimation of acid dissociation constants using graph kernels. *Molecular Informatics, 29*(10), 731–740.
39. Chai, S., et al. (2020). A grand product design model for crystallization solvent design. *Computers & Chemical Engineering, 135*, 106764.
40. Zhang, R., et al. (2025). Artificial intelligence-driven drug toxicity prediction: Advances, challenges, and future directions. *Toxics, 13*(7), 525.
41. Niazi, S. K. (2025). Regulatory perspectives for AI/ML implementation in pharmaceutical GMP environments. *Pharmaceuticals, 18*(6), 901.
42. Huanbutta, K., et al. (2024). Artificial intelligence-driven pharmaceutical industry: A paradigm shift in drug discovery, formulation development, manufacturing, quality control, and post-market surveillance. *European Journal of Pharmaceutical Sciences, 203*, 106938.
43. Aritra, S., Shikha, C. B., & Indu, S. (2025). Harnessing the power of artificial intelligence in pharmaceuticals: Current trends and future prospects. *Intelligent Pharmacy, 3*(3), 181–192.
44. EMA. (2025). *Review of AI/ML applications in medicines lifecycle (2024).*
45. Holey, V. S., & Baitule, A. W. (2024). A wide application of artificial intelligence in pharma field. *Asian Journal of Pharmaceutical Research, 14*(4), 403–410.
46. Vora, L. K., et al. (2023). Artificial intelligence in pharmaceutical technology and drug delivery design. *Pharmaceutics, 15*(7).
47. Kandhare, P., et al. (2025). A review on revolutionizing healthcare technologies with AI and ML applications in pharmaceutical sciences. *Drugs and Drug Candidates, 4*(1), 9.
48. Buiten, M. C. (2019). Towards intelligent regulation of artificial intelligence. *European Journal of Risk Regulation, 10*(1), 41–59.
49. Sanyal, S., Sharma, P., & Dudani, C. (2024). *A complex adaptive system framework to regulate artificial intelligence* (Working Paper).
50. Algarvio, R. C., et al. (2025). Artificial intelligence in pharmacovigilance: A narrative review and practical experience with an expert-defined Bayesian network tool. *International Journal of Clinical Pharmacy, 47*(4), 932–944.
51. Nagar, A., Gobburu, J., & Chakravarty, A. (2025). Artificial intelligence in pharmacovigilance: Advancing drug safety monitoring and regulatory integration. *Therapeutic Advances in Drug Safety, 16*, 20420986251361435.
52. Desai, M. K. (2024). Artificial intelligence in pharmacovigilance – Opportunities and challenges. *Perspectives in Clinical Research, 15*(3), 116–121.
53. Shamim, M. A., et al. (2024). Artificial intelligence and big data for pharmacovigilance and patient safety. *Journal of Medicine, Surgery, and Public Health, 3*, 100139.
54. Bresso, E., et al. (2020). Investigating ADR mechanisms with knowledge graph mining and explainable AI. *arXiv preprint arXiv*, 2012.09077.

55. Jin, M., et al. (2020). *Artificial intelligence in healthcare: Landscape, policies and regulations in Asia-Pacific*. Duke-NUS Medical School.
56. Tsang, L., et al. (2017). The impact of artificial intelligence on medical innovation in the European Union and United States. *Intellectual Property and Technology Law Journal, 29*(8), 3–12.
57. Taeihagh, A. (2021). Governance of artificial intelligence. *Policy and Society, 40*(2), 137–157.
58. Muehlematter, U. J., Daniore, P., & Vokinger, K. N. (2021). Approval of artificial intelligence and machine learning-based medical devices in the USA and Europe (2015–20): A comparative analysis. *The Lancet Digital Health, 3*(3), e195–e203.
59. Van Kolfschooten, H., & Van Oirschot, J. (2024). The EU artificial intelligence act (2024): Implications for healthcare. *Health Policy, 149*, 105152.
60. EU. (2021). Proposal for a regulation of the European Parliament and of the council laying down harmonised rules on Artificial Intelligence (artificial intelligence act) and amending certain union legislative acts. *COM/2021/206final*, 1–107.
61. Busch, F., et al. (2024). Navigating the European Union artificial intelligence act for healthcare. *npj Digital Medicine, 7*(1), 210.
62. MDCG. (2019). *M.D.C.G. Guidance on qualification and classification of software in regulation (EU) 2017/745–MDR and regulation (EU) 2017/746–IVDR*. European Commission.
63. Hines, P. A., et al. (2019). The European Medicines Agency's goals for regulatory science to 2025. *Nature Reviews Drug Discovery, 18*(6), 403–404.
64. Hines, P. A., et al. (2020). Regulatory science to 2025: An analysis of stakeholder responses to the European Medicines Agency's Strategy. *Frontiers in Medicine, 7*, 508.
65. Thakkar, S., et al. (2023). Artificial intelligence and real-world data for drug and food safety – A regulatory science perspective. *Regulatory Toxicology and Pharmacology, 140*, 105388.
66. FDA, U. (2024). *Artificial intelligence & medical products: How CBER, CDER, CDRH, and OCP are working together*. US Food and Drug Administration.
67. FDA. (2019). *Proposed regulatory framework for modifications to artificial intelligence/machine learning (AI/ML)-based software as a medical device (SaMD)*.
68. FDA, U. (2023). *Predetermined change control plans for machine learning-enabled medical devices: Guiding principles*.
69. Health. (2019). *N.I.f. and C. Excellence evidence standards framework for digital health technologies*. NICE.
70. Darzi, A. (2022). *The regulation of artificial intelligence as a medical device*. Regulatory Horizons Council.
71. Ardic, N., & Dinc, R. (2025). Artificial intelligence in healthcare: Current regulatory landscape and future directions. *British Journal of Hospital Medicine, 86*(8), 1–21.
72. Palaniappan, K., Lin, E. Y., & Vogel, S. (2024). Global Regulatory Frameworks for the Use of Artificial Intelligence (AI) in the Healthcare Services Sector. *Healthcare, 12*, 562. https://doi.org/10.3390/healthcare12050562
73. Chen, Z., et al. (2021). Applications of artificial intelligence in drug development using real-world data. *Drug Discovery Today, 26*(5), 1256–1264.
74. Khalid, N., et al. (2023). Privacy-preserving artificial intelligence in healthcare: Techniques and applications. *Computers in Biology and Medicine, 158*, 106848.
75. EU. (2016). *Regulation (EU) 2016/679 of the European Parliament and of the Council*. Official Journal of European Union.
76. Becker, R., Chokoshvili, D., & Dove, E. S. (2024). Legal bases for effective secondary use of health and genetic data in the EU: Time for new legislative solutions to better harmonize data for cross-border sharing? *International Data Privacy Law, 14*(3), 223–246.
77. Shabani, M., & Yilmaz, S. (2022). Lawfulness in secondary use of health data: Interplay between three regulatory frameworks of GDPR, DGA & EHDS. *Technology and Regulation, 2022*, 128–134.
78. Zenker, S., et al. (2022). Data protection-compliant broad consent for secondary use of health care data and human biosamples for (bio)medical research: Towards a new German national standard. *Journal of Biomedical Informatics, 131*, 104096.

79. van Gils-Schmidt, H. J., et al. (2025). Critical consideration towards broad consent by patient experts: Results of a semi-structured interview study on the secondary use of medical data. *BMC Medical Ethics, 26*(1), 163.
80. Kaaya, E. (2024). Biobank consent under the GDPR: Are potential sample donors informed about all lawful uses of biobank data? *Medicine, Health Care and Philosophy, 27*, 567–577.
81. Lenarczyk, G., et al. (2025). The future of AI regulation in drug development: A comparative analysis. *Journal of Law and the Biosciences, 12*(2), lsaf028.
82. Gerke, S., Minssen, T., & Cohen, G. (2020). *Ethical and legal challenges of artificial intelligence-driven healthcare*.
83. Olawade, D. B., et al. (2024). Artificial intelligence in clinical trials: A comprehensive review of opportunities, challenges, and future directions. *International Journal of Medical Informatics, 206*, 106141.
84. Cross, J. L., Choma, M. A., & Onofrey, J. A. (2024). Bias in medical AI: Implications for clinical decision-making. *PLOS Digital Health, 3*(11), e0000651.
85. Hasanzadeh, F., et al. (2025). Bias recognition and mitigation strategies in artificial intelligence healthcare applications. *npj Digital Medicine, 8*(1), 154.
86. Chinta, S. V., et al. (2024). AI-driven healthcare: A review on ensuring fairness and mitigating bias. *arXiv preprint arXiv*, 2407.19655.
87. Ueda, D., et al. (2024). Fairness of artificial intelligence in healthcare: Review and recommendations. *Japanese Journal of Radiology, 42*(1), 3–15.
88. Palaniappan, K., Lin, E. Y. T., & Vogel, S. (2024). Global Regulatory Frameworks for the Use of Artificial Intelligence (AI) in the Healthcare Services Sector. *Healthcare (Basel), 12*(5).
89. Ullagaddi, P., & Ullagaddi, P. (2025). Cross-regional analysis of global AI health-care regulation. *Journal of Computer and Communications, 13*, 66–83.
90. Qiu, Y., et al. (2025). Regulatory sandbox expansion: Exploring the leap from fintech to medical artificial intelligence. *Intelligent Oncology, 1*(2), 120–127.
91. Gerke, S., Minssen, T., & Cohen, G. (2020). Ethical and legal challenges of artificial intelligence-driven healthcare. In *Artificial intelligence in healthcare* (pp. 295–336). Elsevier.

MIX
Papier aus verantwortungsvollen Quellen
Paper from responsible sources
FSC® C105338

If you have any concerns about our products,
you can contact us on
ProductSafety@springernature.com

In case Publisher is established outside the EU,
the EU authorized representative is:
Springer Nature Customer Service Center GmbH
Europaplatz 3, 69115 Heidelberg, Germany

Printed by Libri Plureos GmbH
in Hamburg, Germany